LABORATORY MAN

Botany

LABORATORY MANUAL

Botany

second edition

Darrell S. Vodopich
Baylor University

Randy Moore
University of Akron

WCB McGraw-Hill

Boston, Massachusetts   Burr Ridge, Illinois   Dubuque, Iowa
Madison, Wisconsin   New York, New York   San Francisco, California   St. Louis, Missouri

The McGraw-Hill Companies Higher Education Group
A Division of The McGraw-Hill Companies

Moore/Vodopich: Botany Laboratory Manual

1 2 3 4 5 6 7 8 9 0 QPD QPD 9 0 9 8 7

ISBN 0-697-28628-2

table of contents

preface

We designed this laboratory manual for an introductory botany course. The experiments and procedures are simple, easy to perform, and especially appropriate for large classes. Each exercise includes many photographs, traditional topics, and experiments that work. Procedures within each exercise are numerous and discrete so that an exercise can be tailored to the needs of the students, the style of the instructor, and the facilities available.

To the student. We hope this manual is an interesting guide to many areas of botany. As you survey these areas you'll probably spend equal amounts of time observing and experimenting. Don't hesitate to go beyond the observations that we've outlined—your future success as a scientist depends on your ability to search for and notice things that others may overlook. Now is the time to develop this ability with a mixture of hard work and relaxed observation. Have fun and learning will come easily. Also, remember that this manual is designed with your instructors in mind as well. Go to them often with questions—their experience is a valuable tool that you should use as you work.

To the instructors. The straightforward approach of this manual emphasizes experiments and activities that optimize the students' investment of time and your investment of supplies, equipment, and preparation. Simple and straightforward experiments can be the most effective if you interpret the work in depth. Most experiments can be conducted easily by a student in three hours. Terminology, structures, photographs, and concepts are limited to those the student can readily observe and understand. In each exercise we have included a few activities requiring a greater investment of effort if resources are available, but omitting them will not detract from the objectives.

This manual functions best with an instructor's guidance and is not an auto-tutorial system. We've tried to guide students from observations to conclusions and to make the transition to biological principles. But discussion and interaction between student and instructor are major components of a successful laboratory. Be sure to examine the "Questions for Further Thought and Study" in each exercise. We hope they will help you expand students' perceptions that each exercise has broad applications to their world.

The order of laboratory exercises reflects their treatment in *Botany,* by Moore, et al. However, all exercises are self-contained and compatible with virtually all modern textbooks on general botany. The questions in each exercise have been designed to: direct student observations and experiments, record experimental results, and broaden the context and application of observations.

We've included a section of color art and photographs with detailed captions describing adaptations of plants. We chose this pictorial essay because strategies and adaptations in whole organisms integrate many of the "pieces" of biology investigated in individual exercises. The exercises that survey the kingdoms emphasize continuity. These photographs illustrate diversity and the incredible strategies that plants use to survive and reproduce. Many evolutionary themes can be discussed and based on diversity of adaptations that we've illustrated.

We wish to recognize the staff at WCB/McGraw-Hill and also express our sincere gratitude to Martin Huss, Arkansas State University, and James Runkle, Wright State University, who, in addition to friends and colleagues, contributed to the changes in this second edition.

A *Laboratory Resource Guide* is available from the publisher. It contains helpful comments for instructors on setting up and performing the experiments covered in this laboratory manual. This revised Guide includes expanded descriptions of experimental preparations and specific catalog numbers for purchase of prepared slides and other supplies.

Darrell Vodopich
Randy Moore

1

The Microscope

Objectives

By the end of this exercise you will be able to

1. identify and state the function of the primary parts of a compound microscope and dissecting (stereoscopic) microscope

2. use a compound microscope and dissecting microscope, including being able to (a) carry a microscope properly, (b) focus a microscope, (c) prepare a wet mount, (d) determine the magnification and size of the field of view, and (e) determine the depth of field.

Many organisms and biological structures are too small to be seen with the unaided eye. To observe such specimens botanists often use a **light microscope,** a coordinated system of lenses arranged to produce an enlarged, focusable image of a specimen. A light microscope **magnifies** a specimen, meaning that it increases its apparent size.

Magnification with a light microscope is usually accompanied by improved **resolution,** which is the ability to distinguish two points as separate points. Thus, the better the resolution, the sharper or crisper the image appears. The resolving power of the unaided eye is approximately 0.1 mm (1 in = 25.4 mm), meaning that our eyes can distinguish two points 0.1 mm apart. A good-quality light microscope, used properly, can improve resolution as much as 1,000-fold (i.e., to 0.1 μm).

The ability to discern detail also depends on contrast. Therefore, many specimens examined with a light microscope are stained with artificial dyes that increase contrast and make the specimen more visible.

The invention of the light microscope was profoundly important to biology because it was used to formulate the cell theory and study structure at the cellular level. Today, the light microscope is the most fundamental and important tool of many botanists.

THE COMPOUND LIGHT MICROSCOPE

Study and learn the parts of the typical compound light microscope shown in figure 1.1. A light microscope has two, sometimes three, systems: an illuminating system, an imaging system, and possibly a viewing and recording system.

Illuminating System

The illuminating system, which concentrates light on the specimen, usually consists of a light source, condenser lens, and iris diaphragm. The **light source** is a light bulb located at the base of the microscope. The light source illuminates the specimen by passing light through a thin,

almost transparent part of the specimen. The **condenser lens,** located immediately below the specimen, focuses light from the light source onto the specimen. Just below the condenser is the **iris diaphragm,** which is a knurled ring or lever that can be opened and closed to regulate the amount of light reaching the specimen. When the iris diaphragm is open, the image is bright; when closed, the image is dim.

Imaging System

The imaging system improves resolution and magnifies the image. It consists of the objective and ocular (eyepiece) lenses and a body tube. The **objectives** are three or four lenses mounted on a revolving **nosepiece.** Each objective is

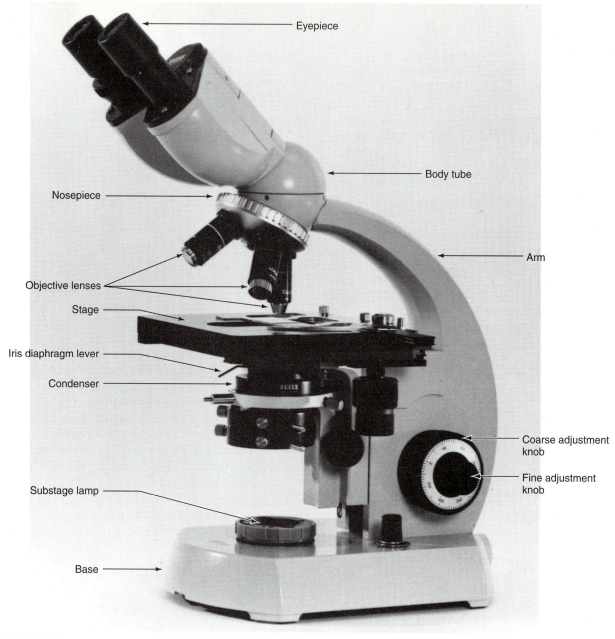

FIGURE 1.1

Major parts of a compound microscope.

actually a series of several lenses that magnify the image, improve resolution, and correct aberrations in the image. The magnifying power of each objective is etched on the side of the lens (e.g., ×4, ×10, or ×40). The **ocular** is the lens that you look through. Microscopes with one ocular are **monocular** microscopes, and those with two are **binocular** microscopes. The oculars usually magnify the image ten times. The **body tube** is a metal casing through which light passes to the oculars. In microscopes with bent body tubes and inclined oculars, the body tube contains mirrors

and a prism that redirects light to the oculars. The **stage** supports and secures the glass slide on which the specimen is mounted.

Viewing and Recording System

The viewing and recording system converts radiation to a viewable and/or permanent image. It usually consists of a camera or video screen. Most student microscopes do not have viewing and recording systems.

1-2

USING A COMPOUND MICROSCOPE

Although the maximum resolving power of light microscopes has not increased significantly during the last century, the construction and design of light microscopes have improved with newer models. For example, built-in light sources have replaced adjustable mirrors in the illuminating system, and lenses are made of better glass than they were in the past. Your lab instructor will review with you the parts of the microscopes (and their functions) you will use in the lab.

After familiarizing yourself with the parts of a microscope, you're now ready for some hands-on experience with the instrument.

Procedure 1.1
Use a compound microscope

1. Remove the microscope from its cabinet and carry it upright with one hand grasping the arm and your other hand supporting the microscope below its base. Place your microscope on the table in front of you.

2. Clean all of the microscope's lenses with lens paper. Do not remove the oculars or any other parts from the body tube of the microscope.

> ### CAUTION
> Do not use paper towels or Kimwipes; they can scratch the lenses.

3. Plug in the microscope and turn on the light source.

4. If it isn't already in position, rotate the nosepiece until the low-power (×4 or ×10) objective is in line with the body tube. You'll feel the objective click into place when it is positioned properly. Always begin examining slides with the low-power objective.

5. Locate the **coarse adjustment** knob on the side of the microscope. Depending on the type of microscope that you're using, the coarse adjustment knob moves either the nosepiece (with its objectives) or the stage to focus the lenses on the specimen. Only a partial turn of the coarse adjustment knob moves the stage or nosepiece a relatively large distance. The coarse adjustment should only be used when you're viewing a specimen under low magnification.

6. Rotate the coarse adjustment knob clockwise to move the objective within 1 cm of the stage (1 cm = 0.4 in). If your microscope is binocular, adjust the distance between the oculars to match your interpupillary distance (i.e., distance between your pupils). If your microscope is monocular, keep both eyes open when using the microscope. After a little practice you will ignore the image received by the eye not looking through the ocular.

7. Place a microscope slide of newsprint of the letter *e* on the horizontal stage so that the *e* is directly below the low-power objective lens and is right side up.

8. Look through the microscope and focus on the *e* by rotating the coarse adjustment knob counterclockwise (i.e., raising the objective lens). If you don't see an image, the *e* is probably off center. Recheck to be sure that the *e* is directly below the objective lens and that you can see a spot of light surrounding the *e*.

9. Adjust the iris diaphragm so that the brightness of the transmitted light provides the best view.

10. Focus up and down to achieve the crispest image.

Question 1

a. As you view the letter *e*, how is it oriented? upside down or right side up? _____

b. How does the image move when the slide is moved to the right or left? _____ up or down? _____

c. What causes the image to move in a different direction from how you move the slide? _____ _____

d. What happens to the brightness of the view when you go from low to high power? _____ _____

e. When you rotate the coarse adjustment knob forward (clockwise), does the specimen get closer to or farther away from the objective? _____ Is this also true for the fine adjustment? _____ Why or why not? _____

Magnification

Procedure 1.2
Determine magnification

1. Estimate the magnification of the *e* by looking at the magnified image and then at the *e* without using the microscope.

2. Examine each objective and record in table 1.1 the magnifications written on the objectives and oculars of your microscope.

TABLE 1.1

TOTAL MAGNIFICATION AND AREAS OF FIELD OF VIEW (FOV) FOR THREE OBJECTIVES

Objective Power	Objective Magnification	Ocular Magnification	Total Magnification	FOV Diameter (mm)	FOV Area (mm²)
Low	×_____	×_____	×_____	_____	_____
Medium	×_____	×_____	×_____	_____	_____
High	×_____	×_____	×_____	_____	_____

3. Calculate and record in table 1.1 the total magnification for each objective following this formula:

$$Mag_{Tot} = Mag_{Obj} \times Mag_{Ocu}$$

Mag_{Tot} = **total magnification of the image**

Mag_{Obj} = **magnification of the objective lens**

Mag_{Ocu} = **magnification of the ocular**

For example, if you're viewing the specimen with a ×4 objective lens and a ×10 ocular, the total magnification of the image is $4 \times 10 = \times 40$; that is, the specimen appears forty times larger than normal.

4. Slowly rotate the high-power (×40) objective into place. Be sure that the objective does not touch the slide. If the objective does not rotate into place without contacting the slide, do not force it; ask your lab instructor to help you.

5. After the high-power objective is in place, notice that the image remains somewhat in focus. Most light microscopes are **parfocal,** meaning that the image remains nearly focused after the high-power objective lens is in place. Most light microscopes are also **parcentered;** the image remains centered after the high-power objective lens is in place.

6. You may need to readjust the iris diaphragm because the high-magnification objective allows less light to pass through to the ocular.

7. To fine-focus the image, locate the **fine adjustment knob** on the side of the microscope. Turning this knob changes the specimen-to-objective distance slightly and therefore makes it easy to fine-focus the image.

8. Compare the size of the image under high magnification with the image under low magnification.

C A U T I O N

Never use the coarse adjustment knob to fine-focus an image on high power.

FIGURE 1.2

The circular, illuminated field of view of a compound microscope (40×).

Question 2

a. How many times is the image of the *e* magnified when viewed through the high-power objective?

_____

b. If you didn't already know what you were looking at, could you determine at high magnification that you were looking at a letter *e*? How? _____

_____

Determining the Size of the Field of View

The **field of view (FOV)** is the area that you can see through the ocular and objective (fig. 1.2). Knowing the size of the field of view is important because you can use it to determine the approximate size of an object you are examining. The field of view can be measured with ruled

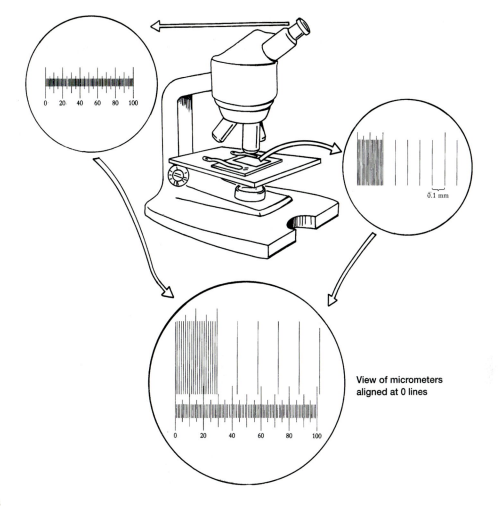

FIGURE 1.3

Stage and ocular micrometers. Micrometers are used to calibrate microscopes and measure the size of specimens.

micrometers (fig. 1.3). An **ocular micrometer** is a small glass disk with uniformly spaced lines etched at unknown intervals. An ocular micrometer is inserted in an ocular, and the distance between its lines is calibrated against a standard ruler called a stage micrometer. A **stage micrometer** is a glass slide with uniformly spaced lines etched at known intervals.

Procedure 1.3
Determine the size of the field of view using ocular and stage micrometers

1. Rotate the ocular until the lines of the ocular micrometer parallel those of the stage micrometer (fig. 1.3).

2. Align lines at the left edges of the two micrometers by moving the stage micrometer (fig. 1.3).

3. Count how many spaces on the stage micrometer fit precisely in a given number of spaces on the ocular micrometer. Record the values below.

$$y \text{ ocular spaces} = x \text{ stage spaces}$$

$$y = \underline{\hspace{1cm}}$$

$$x = \underline{\hspace{1cm}}$$

Since the smallest space on a stage micrometer is 0.01 mm, then

$$y \text{ ocular spaces (mm)} = x \text{ stage spaces} \times 0.01$$

$$1 \text{ ocular space (mm)} = (x/y) \times 0.01$$

4. Calculate the distance in millimeters between lines of the ocular micrometer. For example, if the length of ten spaces on the ocular micrometer equals the length of eight spaces on the stage micrometer, then

$$y = 10$$

$$x = 8$$

10 ocular spaces (mm) = 8 stage spaces × 0.01 mm

1 ocular space (mm) = 8/10 × 0.01 mm

1 ocular space (mm) = 0.008 mm

1 ocular space = 8 μm

Therefore, if a specimen spans eight spaces on your ocular micrometer with that objective in place, that specimen is 64 μm long.

5. Calibrate the ocular micrometer for each objective on your microscope. Record in table 1.1 the diameter of the field of view for each objective.

6. Use this information to determine the area of the circular field of view with the following formula:

area of circle = π × radius² (π = 3.14)

7. Record your calculated field of view areas in table 1.1.

Alternate Procedure 1.3
Use a transparent ruler to determine the size of the field of view

1. Obtain a clear plastic ruler with a metric scale.

2. Place the ruler on the stage of your microscope and rotate the nosepiece to the objective of lowest magnification.

3. Focus with the coarse adjustment and then the fine adjustment until the metric markings on the ruler are clear.

4. Align the ruler to measure the diameter of the circular field of view. The space between each line on the ruler should represent a 1 mm interval.

5. Record in table 1.1 the measured diameter of this low-magnification field of view. Also calculate the radius, which is half the diameter.

6. The ruler cannot be used to measure the diameters of the field of view at medium and high magnifications because the markings are too far apart. Therefore, these diameters must be calculated using the following formula:

FOV$_{low}$ × Mag$_{low}$ = FOV$_{hi}$ × Mag$_{hi}$

FOV$_{low}$ = diameter of the field of the low-power objective

Mag$_{low}$ = magnification of the low-power objective

FOV$_{hi}$ = diameter of the field of view of the high-power objective

Mag$_{hi}$ = magnification of the high-power objective

7. Calculate and record in table 1.1 the diameters of the field of view for the medium and high magnifications.

8. Calculate and record in table 1.1 the circular area of the field of view for the three magnifications by using the following formula:

area of circle = π × radius² (π = 3.14)

Question 3

a. Which provides the largest field of view, the low- or high-power objective? _____

b. How much more area can you see with the ×4 objective than with the ×40 objective? _____

_____

c. Why is it more difficult to locate an object starting with the high-power objective? _____

_____

d. Which objective should you use to initially locate the specimen? Why? _____

_____

Depth of Field

Depth of field is the thickness of the object that is in sharp focus. Depth of field varies with different objectives and magnifications.

Procedure 1.4
Determine the depth of the field of view

1. Using the low-power objective, examine a prepared slide of three colored threads mounted on top of each other.

2. Focus up and down and try to determine the order of the threads from top to bottom. Remember that the order of the threads will not be the same on all slides.

3. Reexamine the threads using the high-power objective lens.

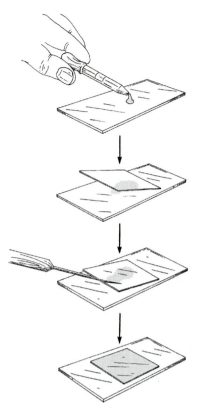

FIGURE 1.4

Preparing a wet mount of a biological specimen.

Question 4

a. Are all three colored threads in focus at low power?

_____

b. Can all three threads be in focus at the same time using the high-power objective? Why or why not?

_____

c. Which objective, high-power or low-power, provides the greatest depth of field? _____

_____

Preparing a Wet Mount of a Biological Specimen

Procedure 1.5
Prepare a wet mount of a biological specimen

1. Place a drop of water from the culture labeled "Algae" on a clean microscope slide.

2. Place the edge of a clean coverslip at one edge of the drop; then slowly lower the coverslip onto the drop so that no air bubbles are trapped (fig. 1.4). Your lab instructor will demonstrate this technique. This fresh preparation is called a **wet mount** and can be viewed with your microscope.

3. Experiment with various intensities of illumination. To do this, rotate the lowest-magnification objective into place and adjust the iris diaphragm to produce the least illumination. Observe the image; note its clarity, contrast, and color. Repeat these observations with at least four different levels of illumination. The fourth level should have the diaphragm completely open.

4. Repeat step 3 for the medium- and high-power objectives.

Question 5

a. Is the image always best with highest illumination?

_____

b. Is the same level of illumination best for all

magnifications? _____

c. Which magnifications require the most illumination

for best clarity and contrast? _____

_____

5. Examine your preparation of algae, and sketch the organisms that you see. Figure 1.5 will help you identify some of these organisms. Don't mistake air bubbles for organisms! Air bubbles appear as uniformly round structures with dark, thick borders.

Question 6

Why is it important to put a coverslip over the drop

of water when preparing a wet mount? _____

_____

Practice

For practice using your microscope, examine the prepared slides available in the lab. You'll examine these slides in more detail in the coming weeks, so don't worry about their contents. Rather, use this exercise to familiarize yourself with the microscope. Also prepare wet mounts of the cultures available in the lab, and sketch the organisms that you see. When you've finished, turn off the light source, and store the microscope in its cabinet.

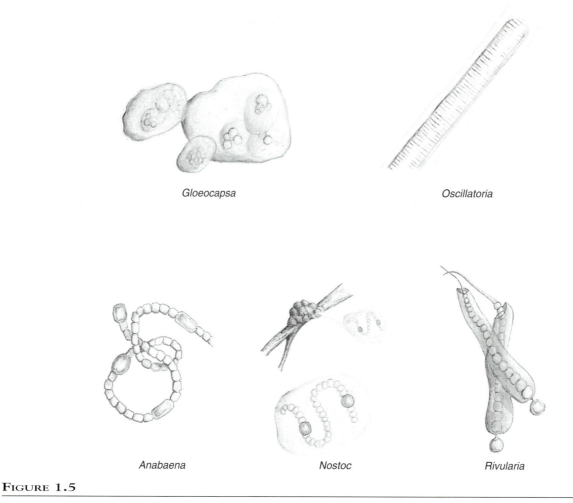

Gloeocapsa

Oscillatoria

Anabaena

Nostoc

Rivularia

FIGURE 1.5

Common cyanobacteria (blue-green algae).

THE DISSECTING (STEREOSCOPIC) MICROSCOPE

A **dissecting (stereoscopic) microscope** offers some advantages over a compound microscope. Although a compound microscope can produce high magnifications and excellent resolution, it has a small **working distance,** which is the distance between the objective lens and specimen. Therefore, it is difficult to manipulate a specimen while observing it with a compound microscope. In contrast, a dissecting microscope is used to view objects that are opaque or too large to see with a compound microscope.

A dissecting microscope provides a much larger working distance. This distance is usually several centimeters (compared to a centimeter or less for a compound microscope), making it possible to dissect and manipulate most specimens. Also, most specimens for dissection are too thick to observe with transmitted light from a light source below the specimen. Therefore, dissecting microscopes use a light source above the specimen; the image is formed from reflected light.

Dissecting microscopes are always binocular (fig. 1.6). Each ocular views the specimen at a different angle through one or more objective lenses. This arrangement provides a three-dimensional image with a large depth of field. This is in contrast to the image in a compound microscope, which is basically two-dimensional. However, the advantages of a stereoscopic microscope are often offset by lower resolution and magnification than a compound microscope. Most dissecting microscopes have magnifications of ×4 to ×50.

Procedure 1.6
Use a dissecting microscope

1. Carry the dissecting microscope to your desk and clean its lenses in the same way you cleaned the lenses of your compound microscope.

2. Use figure 1.6 to familiarize yourself with the parts of your microscope.

3. Use your dissecting microscope to examine the plants available in the lab. Sketch some of these plants.

4. Use a ruler to measure the diameter of the field of view with your dissecting microscope.

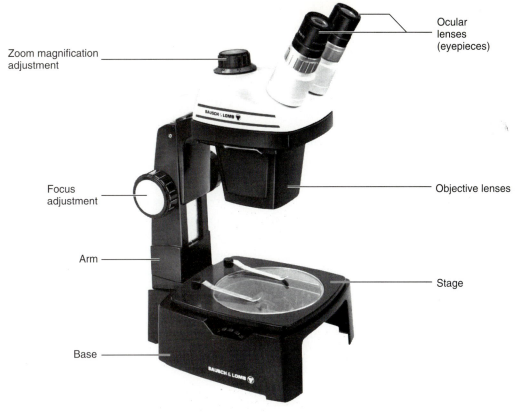

Zoom magnification adjustment

Ocular lenses (eyepieces)

Focus adjustment

Objective lenses

Arm

Stage

Base

FIGURE 1.6

Dissecting (stereoscopic) microscope.

TABLE 1.2

COMPARING COMPOUND AND DISSECTING MICROSCOPES

Characteristic	Compound Microscope	Dissecting Microscope
Magnification		
Resolution		
Size of field of view		
Depth of field		

Question 7

a. What is the area of the field of view when you use the lowest magnification of your dissecting microscope?

_____

b. What is the area when you use the highest

magnification? _____

c. How does the image through a dissecting microscope move when the specimen is moved to the right or

left? _____

up or down? _____

d. How does the direction of illumination differ in dissecting as opposed to compound microscopes?

_____

A COMPARISON OF COMPOUND AND DISSECTING MICROSCOPES

Complete table 1.2, comparing magnification, resolution, size of the field of view, and depth of field of a compound microscope and a dissecting microscope. Use the terms *high*, *low*, or *same* to describe your comparisons.

Summary of Activities

1. Use and examine the parts of a compound microscope.
2. Determine magnification.
3. Measure the size of the field of view.
4. Determine the depth of field.
5. Prepare a wet mount of algae.
6. Use a dissecting microscope.
7. Compare the features of compound and dissecting microscopes.

Questions for Further Thought and Study

1. What is the function of each major part of a compound and a dissecting microscope?
2. What are the advantages of knowing the diameter of the field of view at a given magnification?
3. Why must specimens viewed with the compound microscope be thin? Why are they sometimes stained with dyes?
4. What are the advantages and disadvantages of using a compound microscope as compared to a stereoscopic microscope?
5. Why is depth of field important in studying biological structures?

Writing to Learn Botany

The smallest structures of plant cells are best seen with a transmission electron microscope. Refer to your textbook or other books to describe how an electron microscope can resolve such small structures. Write a short essay about the advantages and limitations of a transmission electron microscope.

Biologically Important Molecules

Objectives

By the end of this exercise you will be able to

1. correctly select and perform tests to detect the presence of carbohydrates, lipids, proteins, and nucleic acids
2. explain the importance of a control in biochemical tests
3. use biochemical tests to identify an unknown compound.

Most organic compounds in living organisms are either carbohydrates, proteins, lipids, or nucleic acids. Each of these macromolecules consists of smaller subunits. These subunits of macromolecules are held together by covalent bonds and have different structures and properties. For example, lipids (made of fatty acids) have many C—H bonds and relatively little oxygen, while proteins (made of amino acids) have amino groups (—NH_2) and carboxyl (—COOH) groups. These characteristic subunits and groups impart different chemical properties to macromolecules—for example, carbohydrates such as glucose are polar and soluble in water, whereas lipids are nonpolar and insoluble in water.

CONTROLLED EXPERIMENTS TO IDENTIFY ORGANIC COMPOUNDS

Scientists have devised several biochemical tests to identify the major types of organic compounds in living organisms. Each of these tests involves two or more treatments: an unknown solution to be identified and controls. As the name implies, an **unknown solution** may or may not contain the substance that the investigator is trying to detect. Only a carefully conducted experiment will reveal its contents. In contrast, **controls** are known solutions. We use controls to validate that our procedure is detecting what we expect it to detect and nothing more. During the experiment we compare the unknown solution's response to the experimental procedure with the control's response to the same procedure.

A **positive control** contains the variable for which you are testing; it reacts positively and demonstrates the test's ability to detect what you expect. For example, if you are testing for protein in unknown solutions, then an appropriate positive control is a known solution of protein. A positive reaction shows that your test reacts correctly; it also shows you what a positive test looks like.

A **negative control** does not contain the variable for which you are searching. It contains only the solvent (often distilled water with no solute) and does not react in the test. A negative control shows you what a negative result looks like.

Controls are important because they reveal the specificity of a particular test. For example, if water and a glucose solution react similarly in a particular test, the test cannot distinguish water from glucose. But if the glucose solution reacts differently from distilled water, then the test can distinguish water from glucose. In this instance, the distilled water is a negative control for the test and a known glucose solution is a positive control.

CARBOHYDRATES

Benedict's Test for Reducing Sugars

Carbohydrates are molecules made of carbon (C), hydrogen (H), and oxygen (O) in a ratio of 1:2:1 (e.g., the chemical formula for glucose is $C_6H_{12}O_6$). Carbohydrates are made of **monosaccharides,** or simple sugars (fig. 2.1). Paired

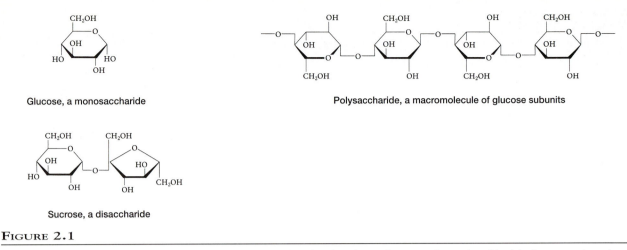

Glucose, a monosaccharide

Polysaccharide, a macromolecule of glucose subunits

Sucrose, a disaccharide

FIGURE 2.1

Carbohydrate macromolecules are made of subunits of mono- or disaccharides. These subunits can be combined by dehydration synthesis (see Fig. 7.3) to form polysaccharides.

monosaccharides form **disaccharides.** Sucrose, for example, is a disaccharide of glucose linked to fructose. Similarly, linking three or more monosaccharides forms a **polysaccharide** such as starch, glycogen, or cellulose (see plate III, fig. 6).

Question 1

Examine the parts of figure 2.1. Shade with a pencil the reactive groups of the glucose molecule that are involved in forming a polysaccharide.

Linkage of subunits in carbohydrates, as well as other macromolecules, often involves the removal of a water molecule (dehydration). Examine figures 7.1, 7.2, and 7.3 to become familiar with the mechanism of synthesis and hydrolysis in carbohydrates.

Many monosaccharides such as glucose and fructose are **reducing sugars,** meaning that they possess free aldehyde —CHO or ketone $-\overset{|}{C}=O$ groups that reduce weak oxidizing agents such as the copper in **Benedict's reagent.** Benedict's reagent contains cupric (copper) ion complexed with citrate in alkaline solution. Benedict's test identifies reducing sugars based on their ability to reduce the cupric (Cu^{2+}) ions to cuprous oxide at basic (high) pH. Cuprous oxide is green to reddish orange.

Oxidized Benedict's reagent (Cu^{2+}) + Reducing sugar (R—COH)

(BLUE)

↓ **Heat**
↓ **High pH**

Reduced Benedict's reagent (Cu^+) + Oxidized sugar (R—COOH)

(GREEN TO REDDISH ORANGE)

A green solution indicates a small amount of reducing sugars, and reddish orange indicates an abundance of reducing sugars. Nonreducing sugars such as sucrose produce no change in color (i.e., the solution remains blue).

Procedure 2.1
Perform Benedict's test for reducing sugars

1. Obtain seven test tubes and number them 1 to 7.
2. Add to each tube the materials to be tested (see table 2.1). Your instructor may ask you to test some additional materials. If so, include additional numbered test tubes. Add 2 mL of Benedict's solution to each tube.
3. Place all of the tubes in a boiling water bath for 3 minutes and observe color changes during this time.
4. After 3 minutes, remove the tubes from the water bath and let them cool to room temperature. Record the color of their contents in table 2.1.

Question 2

a. Which is a reducing sugar, sucrose or glucose? _____

_____

b. Which contains more reducing sugars, potato juice

or onion juice? _____

c. What does this tell you about how sugars are stored

in onions and potatoes? _____

_____

d. Which of the solutions is a positive control? _____

a negative control? _____

Iodine Test for Starch

Staining by iodine (iodine-potassium iodide, I_2KI) distinguishes starch from monosaccharides, disaccharides, and other polysaccharides. The basis for this test is that starch is a coiled polymer of glucose; iodine interacts with these coiled molecules and becomes bluish black. Iodine does not react with carbohydrates that are not coiled and remains yellowish brown. Therefore, a bluish black color is a positive test for starch, and a yellowish brown color (i.e.,

TABLE 2.1

SOLUTIONS AND COLOR REACTIONS FOR (1) BENEDICT'S TEST FOR REDUCING SUGARS AND (2) IODINE TEST FOR STARCH

Tube	Solution	Benedict's Color Reaction	Iodine Color Reaction
1	10 drops onion juice		
2	10 drops potato juice		
3	10 drops sucrose solution		
4	10 drops glucose solution		
5	10 drops distilled water		
6	10 drops reducing-sugar solution		
7	10 drops starch solution		
8			
9			

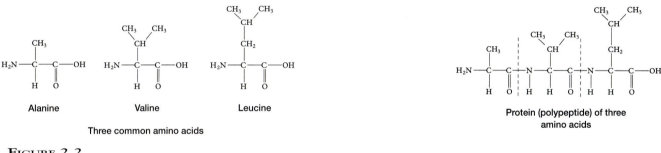

Alanine   Valine   Leucine

Three common amino acids

Protein (polypeptide) of three amino acids

FIGURE 2.2

Proteins are macromolecules with subunits of amino acids. A peptide bond (dotted line) links each amino acid. Each amino acid has one carbon that is bonded to both an amine group (H_2N—) and a carboxyl group (—COOH).

no color change) is a negative test for starch. Notably, glycogen, a common polysaccharide in animals, has a slightly different structure than does starch and produces only an intermediate color reaction.

Procedure 2.2
Perform iodine test for starch

1. Obtain seven test tubes and number them 1 to 7.
2. Add to each tube the materials to be tested (see table 2.1). Your instructor may ask you to test some additional materials. If so, include additional numbered test tubes.
3. Add 3 to 5 drops of iodine to each tube.
4. Record the color of the tubes' contents in table 2.1.

Question 3
a. Which colors more intensely, onion juice or potato juice? _____

b. What does this tell you about how these plants store carbohydrates? _____

c. Which of the solutions is a positive control?

_____ a negative control?

_____

PROTEINS

Proteins are made of amino acids, each of which has an amino group (—NH_2) and a carboxyl (acid) group (—COOH) (see plate III, fig. 5). The bond between these two groups found on adjacent amino acids in a protein is a **peptide bond** (fig. 2.2) and is identified by a **Biuret test.** Specifically, peptide bonds (C—N bonds) in proteins complex with Cu^{2+} in Biuret reagent and produce a violet color. A Cu^{2+} must combine with four to six peptide bonds to produce a color; therefore, free amino acids do not react positively. Long-chain polypeptides (proteins) have many peptide bonds and produce a positive reaction.

Biuret reagent is a 1% solution of $CuSO_4$ (copper sulfate). A violet color is a positive test for the presence of protein; the intensity of color relates to the number of peptide bonds reacting.

TABLE 2.2

SOLUTIONS AND COLOR REACTIONS FOR THE BIURET TEST FOR PROTEIN

Tube	Solution	Color
1	2 mL egg albumin	
2	2 mL honey	
3	2 mL amino acid solution	
4	2 mL distilled water	
5	2 mL protein solution	
6		
7		

Question 4

Examine figure 2.2. Shade with a pencil the reactive amino and carboxyl groups on the three common amino acids shown.

Procedure 2.3
Perform Biuret test for protein

1. Obtain five test tubes and number them 1 to 5. Your instructor may ask you to test some additional materials. If so, include additional numbered test tubes.
2. Add the materials listed in table 2.2.
3. Add 2 mL of 2.5% sodium hydroxide (NaOH) to each tube.

> ### CAUTION
> Do not spill the NaOH. It is extremely caustic.

4. Add 3 drops of Biuret reagent to each tube and mix.
5. Record the color of the tubes' contents in table 2.2.

Question 5

a. Which contains more protein (C—N bonds), egg albumen or honey? _____

b. Do free amino acids have peptide bonds? _____

LIPIDS

Lipids include a variety of molecules that dissolve in nonpolar solvents such as ether and acetone but do not dissolve in polar solvents such as water. Triglycerides are abundant lipids that are made of glycerol and three fatty acids (fig. 2.3). Tests for lipids are based on a lipid's ability to selectively absorb pigments in fat-soluble dyes such as Sudan IV.

Question 6

Examine figure 2.3. Shade with a pencil the reactive groups on the glycerol and fatty acid molecules involved in forming a triglyceride lipid.

Procedure 2.4
Perform the Sudan IV test for lipid

1. Obtain five test tubes and number them 1 to 5. Your instructor may ask you to test some additional materials. If so, include additional numbered test tubes.
2. Add the materials listed in table 2.3.
3. Add 3 mL of water to each tube.
4. Add 5 drops of water to tube 1 and 5 drops of Sudan IV to each of the remaining tubes. Mix the contents of each tube. Record the color of the tubes' contents in table 2.3.

Question 7

a. Is salad oil soluble in water? _____

b. Compare tubes 1 and 2. What is the distribution of the dye with respect to the separated water and oil?

_____

c. What observation indicates a positive test for lipid?

_____

d. Does honey contain much lipid? _____

e. Lipids supply more than twice as many calories per gram as do carbohydrates. Based on your results, which contains more calories per unit serving, oil or honey? _____

_____

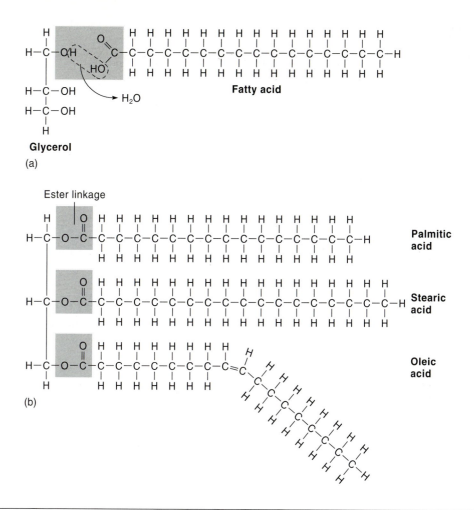

FIGURE 2.3

The structure of a fat. (*a*) An ester linkage forms when the carboxyl group of a fatty acid links to the hydroxyl group of glycerol, with the removal of a water molecule. (*b*) Fats are triacylglycerides whose fatty acids vary in length and also in the presence and location of carbon—carbon double bonds.

TABLE 2.3

SOLUTIONS AND COLOR REACTIONS FOR THE SUDAN IV TEST FOR LIPIDS

Tube	Solution	Description of the Reaction
1	1 mL salad oil + water	
2	1 mL salad oil + Sudan IV	
3	1 mL honey + Sudan IV	
4	1 mL distilled water + Sudan IV	
5	1 mL known lipid solution + Sudan IV	
6		
7		

TABLE 2.4

MATERIALS AND GREASE-SPOT REACTIONS AS A TEST FOR LIPID CONTENT

Food Product	Description of Grease-Spot Reaction
1	
2	
3	
4	
5	
6	

Grease-Spot Test for Lipids

A simpler test for lipids is based on their ability to produce translucent grease marks on unglazed paper.

Procedure 2.5
Perform grease-spot test for lipids

1. Obtain a piece of brown wrapping paper from your lab instructor.
2. Use an eyedropper to add a drop of salad oil near one corner of the piece of paper.
3. Add a drop of water near the opposite corner of the paper.
4. Let the fluids evaporate.
5. Look at the paper as you hold it up to a light.
6. Test other food products and solutions available in the lab in a similar way and record your results in table 2.4.

Question 8
Which of the food products that you tested contain large amounts of lipid?

NUCLEIC ACIDS

DNA and RNA are **nucleic acids** made of nucleotide sub-units (fig. 2.4; see plate II, fig. 4). One major difference between DNA and RNA is their sugar; DNA contains deoxyribose, whereas RNA contains ribose. DNA can be identified chemically with the **Dische diphenylamine test.** Acidic conditions convert deoxyribose to a molecule that binds with diphenylamine to form a blue complex. The intensity of the blue color is proportional to the concentration of DNA.

Question 9
Examine figure 2.4. Shade with a pencil the group on ribose and deoxyribose that reacts upon combining with the phosphate when forming a nucleotide. Also shade the reactive groups that combine with a nitrogenous base.

Procedure 2.6
Perform Dische diphenylamine test for DNA

1. Obtain four test tubes and number them 1 to 4. Your instructor may ask you to test some additional materials. If so, include additional numbered test tubes.
2. Add the materials listed in table 2.5.
3. Add 2 mL of the Dische diphenylamine reagent to each tube and mix thoroughly.
4. Place the tubes in a boiling water bath to speed the reaction.
5. After 10 minutes, transfer the tubes to an ice bath. Gently mix and observe the color of their contents as the tubes cool. Record your observations in table 2.5.

Question 10

a. How does the color compare between tubes 1 and 2?

_____

Why? _____

b. Do DNA and RNA react alike? _____

Why or why not? _____

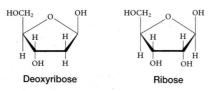

Deoxyribose   Ribose

The two sugar subunits of nucleic acids

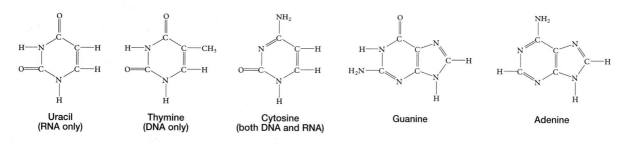

Uracil
(RNA only)

Thymine
(DNA only)

Cytosine
(both DNA and RNA)

Guanine

Adenine

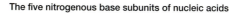

The five nitrogenous base subunits of nucleic acids

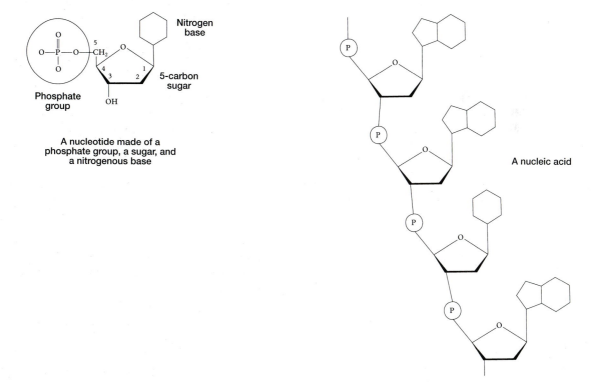

A nucleotide made of a
phosphate group, a sugar, and
a nitrogenous base

A nucleic acid

FIGURE 2.4

Nucleic acids are made of linked nucleotides. Deoxyribonucleic acid (DNA) includes the sugar deoxyribose combined with phosphate groups and combinations of thymine, cytosine, guanine, and adenine. Ribonucleic acid (RNA) includes the sugar ribose with phosphate groups and combinations of uracil, cytosine, guanine, and adenine.

TABLE 2.5

SOLUTIONS AND COLOR REACTIONS FOR DISCHE DIPHENYLAMINE TEST FOR DNA

Tube	Solution	Color
1	2 mL DNA solution	
2	1 mL DNA solution + 1 mL water	
3	2 mL RNA solution	
4	2 mL distilled water	
5		
6		

IDENTIFYING UNKNOWN COMPOUNDS

Each of the previously described tests is relatively specific; that is, iodine produces a bluish black color with starch but not with protein, lipid, nucleic acids, or other carbohydrates. This specificity can be used to identify the contents of an unknown solution.

Procedure 2.7
Identifying unknown solutions

1. Obtain an unknown solution from your laboratory instructor. Record its number in table 2.6.
2. Obtain ten clean test tubes.
3. Number five tubes for the sample as S1 to S5. Number the other five tubes as controls C1 to C5.
4. Place 2 mL of your unknown solution into each of tubes S1 to S5.
5. Place 2 mL of distilled water into each of tubes C1 to C5.
6. Use the five procedures described previously and listed in table 2.6 to detect reducing sugar, starch, protein, lipid, and DNA in your unknown. Your unknown may contain one, none, or several of these macromolecules. Record your results in table 2.6. Submit table 2.6 and the following report to your instructor before you leave the lab.

Summary of Activities

1. Test for reducing sugars using Benedict's test.
2. Test for starch using iodine.
3. Test for protein using the Biuret test.
4. Test for lipid using Sudan IV.
5. Test for DNA using the Dische diphenylamine test.
6. Determine the contents of an unknown solution.

Questions for Further Thought and Study

1. Why did you include controls in all of your tests?
2. Are controls always necessary? Why or why not?
3. How would you determine in what form a plant stores its carbohydrates?

Doing Botany Yourself

Design a procedure to indicate the amount of starch present in various plant tissue samples. Would you weigh your samples? Would you treat them in any way? How would you quantify the iodine test?

Writing to Learn Botany

What are the limitations of these common techniques in detecting the presence of a class of macromolecules? Do botanists who study starch in plant cells commonly use the iodine test for starch? Why or why not?

TABLE 2.6

CHEMICAL TESTING TO IDENTIFY AN UNKNOWN SOLUTION   UNKNOWN NO. _____

Biochemical Test	Color		Conclusion
	Sample	Control	(+/–)
Benedict's Reagent (Reducing Sugars)			
Iodine (Starch)			
Biuret Reagent (Protein)			
Sudan IV (Lipid)			
Dische Diphenylamine Reagent (DNA)			

Report: Identity of Unknown

Indicate which of the following are in your unknown:

Reducing sugars

Starch

Protein

Lipid

DNA

Name: _____   Unknown Number: _____

Lab Section: _____

Separating Organic Compounds

_3_

Objectives

By the end of this exercise you will be able to

1. describe the basis for column chromatography, paper chromatography, and gel electrophoresis
2. use column chromatography and gel electrophoresis to isolate organic compounds from mixtures.

Plant cells are a mixture of the kinds of organic compounds that you studied in the previous exercise: proteins, carbohydrates, lipids, and nucleic acids. All of these compounds are important to life. To understand how plants function, biologists characterize and study these

compounds. This requires that biologists isolate the compounds, such as amino acids and nucleotides, from mixtures. Three techniques that biologists use to isolate compounds from mixtures are column chromatography, paper chromatography, and gel electrophoresis.

In today's lab you will use column chromatography, paper chromatography, and gel electrophoresis to separate compounds from mixtures. Although the procedures you'll learn are simple and may not involve cellular extracts, they model how these techniques are used by biologists in their research.

COLUMN CHROMATOGRAPHY

Column chromatography separates molecules according to their size and shape. The procedure is simple and involves placing a sample onto a column of beads with tiny pores. There are two ways that molecules can move through the column of beads: a fast route between the beads or a slower route through the tiny pores of the beads. Molecules too big to fit into the beads' pores move through the column quickly, while smaller molecules enter the beads' pores and move through the column more slowly (fig. 3.1). The movement of the molecules is analogous to going through or walking around a maze: it takes more time to walk through a maze than to walk around it.

The apparatus used for column chromatography is shown in figure 3.2 and consists of a chromatography column, a matrix, and a buffer.

- The **chromatography column** is a tube with a frit and a spout at its bottom. The **frit** is a membrane or

porous disk that supports and keeps the matrix in the column but allows water and solutes to pass.

- The **matrix** is the material in the column that fractionates, or separates, the chemicals mixed in the sample. The matrix consists of beads with tiny pores and internal channels. The size of the beads' pores determines the matrix's **fractionation range,** which is the range of molecular weights the matrix can separate. Different kinds of matrices have different fractionation ranges. In today's lab, you'll use a matrix with a fractionation range of 1,000 to 5,000.

- The **buffer** helps control the pH of the sample. A buffer is a solution with a known pH that resists changes in pH if other chemicals are added. The pH of a buffer remains relatively constant. This is important because molecules such as proteins often have different shapes at different pH. The buffer carries the sample through the matrix, which separates the chemicals mixed in the sample.

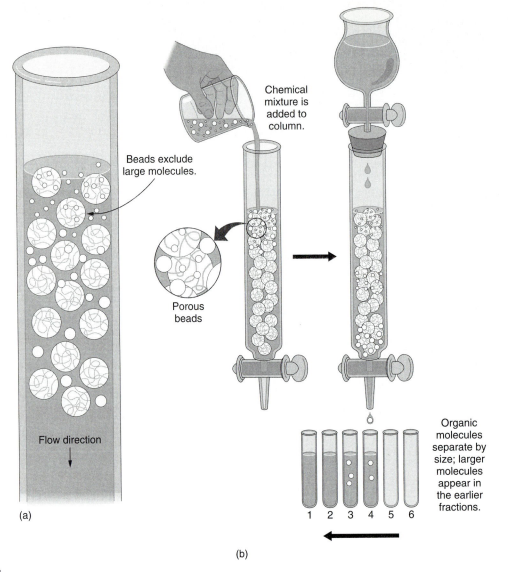

Chemical
mixture is
added to
column.

Beads exclude
large molecules.

Porous
beads

Flow direction

(a)

Organic
molecules
separate by
size; larger
molecules
appear in
the earlier
fractions.

1 2 3 4 5 6

(b)

FIGURE 3.1

Separation of organic molecules by column chromatography. (*a*) As the solution flows through the column, the smaller molecules are retarded as they pass through the canals of the porous beads. Medium-sized molecules will pass through a bead with canals less frequently, and the largest molecules will quickly flow around all the beads. (*b*) The exiting fluid is collected in fractions. The first fractions collected will contain the largest molecules.

As they move through the matrix, small molecules spend much time in the maze of channels and pores in the matrix. Large molecules do not fit into the channels and pores and therefore move around the beads of the matrix. Consequently, larger molecules move through the matrix faster than do smaller molecules.

Column chromatography can also separate compounds of the same molecular weight but different shapes. Compact, spherical molecules penetrate the pores and channels of the matrix more readily than do rod-shaped molecules. Thus, spherical molecules move through a column more slowly than do rod-shaped molecules.

In column chromatography, the buffer containing the sample mixture of chemicals moves through the column and is collected sequentially in test tubes from the bottom of the column. Botanists then assay the content of the tubes to determine which tubes contain the compounds they're interested in.

Question 1

In today's lab, you'll isolate colored compounds from mixtures. However, most biological samples are colorless. How would you determine the contents of the test tubes if all of the samples were transparent?

_____

_____

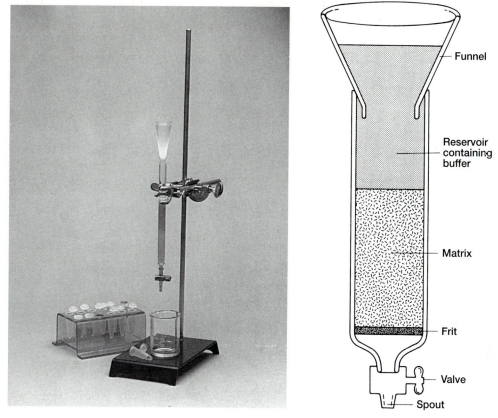

FIGURE 3.2

Apparatus for column chromatography.

Procedure 3.1

Separate compounds by column chromatography

1. Label nine microtubes 1 to 9.

2. Obtain an apparatus for column chromatography and carefully remove all of the buffer from above the bed with a transfer pipet. Do not remove any of the matrix.

3. Obtain a sample to be separated. The sample is a mixture of Orange G (molecular weight = 452) and dextrose, a rodlike polymer of glucose stained blue and having a molecular weight of about 2,000,000.

4. Use a transfer pipet to slowly load 0.2 mL of the sample onto the top of the bed. Drip the sample down the inside walls of the column.

5. Place a beaker under the column.

6. Open the valve. This causes the sample to enter the bed. Close the valve after the sample has completely entered the bed (i.e., when the top of the bed is exposed to air).

7. Use a transfer pipet to slowly cover the bed with buffer. Add buffer until the reservoir is almost full.

8. Hold microtube 1 under the tube and open the valve until you've collected about 1.0 mL of liquid.

9. Repeat step 8 for tubes 2 to 9. The sample will separate in the column.

10. Identify the tubes containing (a) the most orange dye and (b) the most blue dye that eluted from the column.

11. Refill the reservoir with buffer and cover the reservoir with Parafilm.

Question 2

a. Explain your results. Was the color separation distinctive? _____ Would you expect a longer column to more clearly separate the compounds? _____

   Why or why not? _____

b. Suppose your sample had consisted of a mixture of compounds having molecular weights of 50,000, 100,000, and 1,000,000. What kind of results would you predict? _____ Explain your answer. _____

TABLE 3.1

CHROMATOGRAPHY DATA FOR DETERMINING AMINO ACID UNKNOWNS

Tick Mark Number	Amino Acid or Sample Number	Distance to Solvent Front	Distance Traveled by Sample	R$_f$	Identity of Unknown
1					
2					
3					
4					
5					

PAPER CHROMATOGRAPHY FOR SEPARATING AMINO ACIDS

Botanists often analyze the amino acid content of samples to determine protein sequences and enzyme structures. Amino acids can be separated by partitioning them between the stationary and mobile phases of **paper chromatography.** The stationary phase is the paper fibers, and the mobile phase is an organic solvent that moves along the paper. Separation by chromatography begins by applying a liquid sample to a small spot on an origin line at one end of a piece of chromatography paper. The edge of the paper is then placed in a solvent. As the solvent moves up the paper, any sample molecules that are soluble in the solvent will move with the solvent. However, some molecules move faster than others based on their solubility in the mobile phase and their attraction to the stationary phase. These two competing factors are different for different molecular structures, so each type of molecule moves at a different speed and occurs at a different position on the finished **chromatogram.**

Amino acids in solution have no color but react readily with molecules of **ninhydrin** to form a colored product. A completed chromatogram is sprayed with a ninhydrin solution and heated to detect the amino acid spots. The distance of these spots from the origin is measured and used to quantify the movement of a sample. The resulting R$_f$ value (retardation factor) characterizes a known molecule in a known solvent under known conditions and is calculated as follows:

$$R_f = \frac{\text{Distance moved by sample}}{\text{Distance from origin to solvent front}}$$

Procedure 3.2

Separate amino acids and identify unknowns by paper chromatography

1. Obtain a piece of chromatography paper 15 cm square. Avoid touching the paper with your fingers. Use gloves, tissue, or some other means to handle the paper.
2. Lay the paper on a clean paper towel. Then draw a light pencil line 2 cm from the bottom edge of the paper.
3. Draw five tick marks at 2.5 cm intervals from the left end of the line. Lightly label the marks 1 to 5 below the line.
4. Locate the five solutions available for the chromatography procedure. Three of the solutions are known amino acids. One solution is an unknown. The last solution is a plant extract or another unknown.
5. Use a wooden applicator stick to "spot" one of the solutions on mark 1. To do this, dip the stick in the solution and touch it to the paper to apply a small drop (2–3 mm). Let the spot dry and make three more applications. Dry between each application. Record in table 3.1 the name of the solution next to the appropriate mark number.
6. Repeat step 5 for each of the other solutions.
7. Staple or paperclip the edges of the paper to form a cylinder with the spots on the outside and at the bottom.
8. Obtain a quart jar containing the chromatography solvent. The solvent should be 1 cm or less deep. The solvent consists of butanol, acetic acid, and water (2:1:1).
9. Place the cylinder upright in the jar (fig. 3.3). The solvent will be below the pencil line and marks. Close the lid to seal the jar.

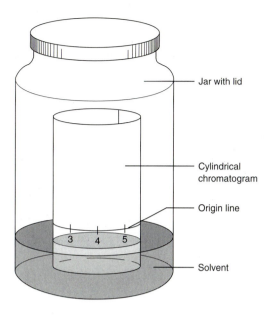

Jar with lid

Cylindrical chromatogram

Origin line

3 4 5

Solvent

FIGURE 3.3

Apparatus for paper chromatography. Numbers on the chromatogram indicate the positions of multiple samples.

10. Keep the jar out of direct light and heat. Allow the solvent to move up the paper for 2 hours, but not all the way to the top.

11. Open the jar and remove the chromatogram. Unclip and flatten the paper. Dry it with a fan or hair dryer. Work under a hood if possible to avoid breathing the solvent vapors.

12. Spray the chromatogram with ninhydrin. Carefully dry the chromatogram with warm air.

13. Circle with a pencil each of the spots. Measure the distance each of the spots has traveled and calculate the R_f for each spot. Record the values in table 3.1.

14. Determine the contents of the unknown solutions by comparing R_f values. Record the results in table 3.1.

GEL ELECTROPHORESIS

Gel electrophoresis separates molecules according to their charge, shape, and size. Buffered samples (mixtures of organic chemicals) are loaded into a Jell–O-like gel, after which an electrical current is placed across the gel. This current moves the charged molecules toward either the cathode or anode of the electrophoresis apparatus. The speed, direction, and distance that each molecule moves are related to its charge, shape, and size.

The apparatus for gel electrophoresis, shown in figure 3.4, consists of an electrophoresis chamber, gel, buffer, samples, and a power supply.

• The **gel** is made by dissolving agarose powder (a derivative of agar) in hot buffer. When the solution cools, it solidifies into a gel with many pores that function as a molecular sieve. The gel is submerged in a buffer-filled chamber containing electrodes.

• The **buffer** conducts electricity and helps control the pH. The pH affects the stability and charge of the samples.

• The **samples** are mixtures of chemicals that are loaded into wells in the gel. They move in the gel during electrophoresis. Samples are often mixed with glycerol or sucrose to make them denser than the buffer so that they will not mix with the buffer.

• The **power supply** provides a direct current across the gel. Charged molecules respond to the current by moving from the sample wells into the gel. Negatively charged molecules move toward the positive electrode (anode), while positively charged molecules move toward the negative electrode (cathode). The greater the voltage, the faster the molecules move.

The sieve properties of the gel affect the rate of movement of the sample into the gel. Small molecules move more easily through the pores than do larger molecules. Consequently, small, compact (e.g., spherical) molecules move faster than do large, rodlike molecules. If molecules have similar shapes and molecular weights, the particles with the greatest charge move fastest and, therefore, farthest.

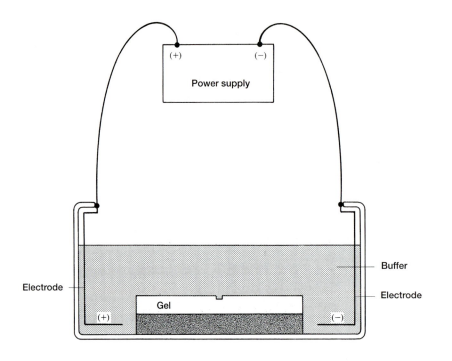

FIGURE 3.4

Apparatus for gel electrophoresis. Gel electrophoresis separates molecules according to their charge, shape, and size.

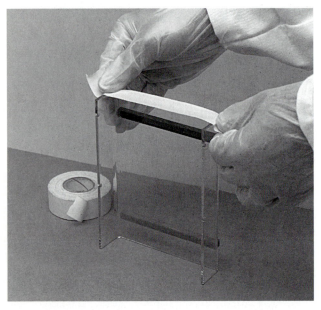

FIGURE 3.5

Tape the ends of the removable gel-bed.

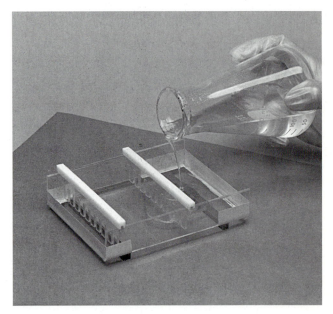

FIGURE 3.6

Prepare the agarose gel solution and cast the gel.

Procedure 3.3
Perform gel electrophoresis

1. Obtain an electrophoresis chamber. Tape the ends of the bed as shown in figure 3.5 and as demonstrated by your instructor.

2. Place a six-tooth comb in the middle set of notches of the gel-cast bed. There should be a small space between the bottom of the teeth and the bed.

3. Mix a 0.8% (weight by volume) mixture of agarose powder in a sufficient volume of buffer to fill the gel chamber. Heat the mixture until the agarose dissolves.

4. When the hot agarose solution has cooled to 50°C, pour the agarose solution into the gel-cast bed (fig. 3.6).

5. After the gel has solidified, gently remove the comb by pulling it straight up (fig. 3.7). Use the sketch in figure 3.8 to label the wells formed in the gel by the comb.

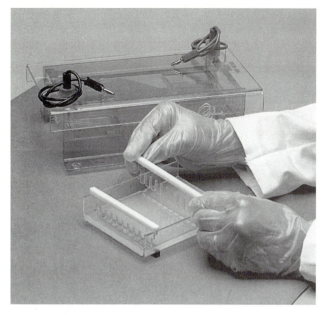

FIGURE 3.7

After the gel has solidified, remove the tape and pull the combs straight up from the gel.

6. Submerse the gel under the buffer in the electrophoresis chamber.

7. You will study six samples:

sample 1: bromophenol blue (molecular weight = 670)

sample 2: methylene blue (molecular weight = 320)

sample 3: Orange G (molecular weight = 452)

sample 4: xylene cyanol (molecular weight = 555)

samples 5 and 6: unknowns

Gently squeeze the pipet bulb to draw sample 1 into the pipet. Be sure that the sample is in the lower part of the pipet. If the sample becomes lodged in the bulb, tap the pipet until the sample moves into the lower part of the pipet.

8. Hold the pipet above the sample tube and slowly squeeze the bulb until the sample is near the pipet's opening.

9. Place the pipet tip into the electrophoresis buffer so it is barely inside sample well 1 (fig. 3.9). Do not touch the bottom of the sample well. Maintain pressure on the pipet bulb to avoid pulling buffer into the pipet.

10. Slowly inject the sample into the sample well. Stop squeezing the pipet when the well is full.

11. Thoroughly rinse the pipet with distilled water.

12. Load the remaining five samples into the gel by repeating steps 7 to 11 (fig. 3.9). Load sample 2 into the second well, sample 3 into the third well, etc.

FIGURE 3.8

Sketch of the wells formed in the gel by the comb.

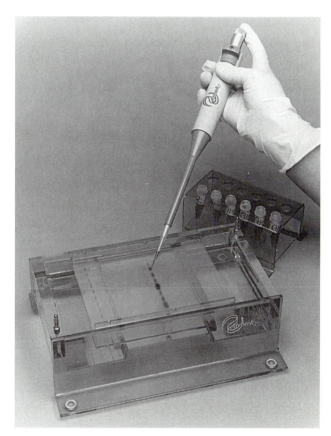

FIGURE 3.9

Submerge the gel in the buffer-filled electrophoresis chamber and load the samples into the wells of the gel.

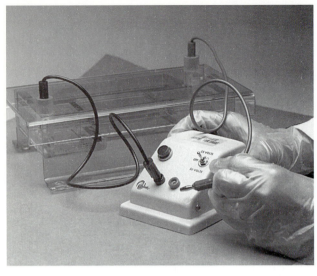

FIGURE 3.10

Attach the safety cover, connect the power source, and run the electrophoresis.

13. Carefully snap on the cover of the electrophoresis chamber (fig. 3.10). The red plug in the cover should be placed on the terminal indicated by the red dot. The black plug in the cover should be placed on the terminal indicated by the black dot.

14. Insert the plug of the black wire into the black (negative) input of the power supply. Insert the plug of the red wire into the red (positive) input of the power supply.

15. Turn on the power and set the voltage at 50 V. You'll soon see bubbles forming on the electrodes. Examine the gel every 10 minutes.

16. After 40 minutes, turn off the power and disconnect the leads from the power source. Gently remove the cover from the chamber and sketch your results in figure 3.8.

Question 3

a. Orange G, bromophenol blue, and xylene cyanol each have a negative charge at neutral pH, while methylene blue has a positive charge at neutral pH. How does this information relate to your results?

_____

b. Did Orange G, bromophenol blue, and xylene cyanol move the same distance in the gel? _____

Why or why not? _____

c. What compounds do you suspect are in samples 5 and 6? _____

Explain your answer. _____

_____

INTERPRETING A DNA-SEQUENCING GEL

Examine figure 3.11, which includes a photograph of a gel used to determine the order, or sequence, of nucleotides in a strand of DNA. To prepare the sample for electrophoresis, samples of the DNA being investigated were put into each of four tubes and induced to replicate. Also, into the first tube an adenine-terminator was added in addition to all the other nucleotides. As the complementary strand was being constructed the terminators were occasionally incorporated wherever an adenine nucleotide was used. This random incorporation resulted in all possible lengths of DNA pieces that had an adenine on the end. The same process was conducted in the other tubes with thymine-, guanine-, and cytosine-terminators—one treatment for each of the four lanes in the gel. Electrophoresis separated the replicated pieces of DNA by size. Staining the gel revealed which lengths of the complimentary DNA were terminated by which nucleotide-terminators.

The gel consists of four "lanes," labeled A, T, G, and C, indicating either adenine-, thymine-, guanine-, or cytosine-terminated pieces of DNA. By "reading" down the gel, you can determine the sequence of nucleotides in the DNA. For example, the uppermost band of the gel is in the C (cytosine) lane. Therefore, the first base of the piece of DNA is cytosine. Similarly, the next two bands are also in the C lane, after which the next bands are in the T, T, G, and C lanes. Thus, the first seven bases of the complementary strand DNA are C-C-C-T-T-G-C. List the next twenty nucleotides of the DNA as indicated by the gel. Also list the sequence of the first twenty-seven nucleotides in the original DNA being investigated.

Question 4

a. How did the sequence of nucleotides revealed on the gel differ from the sequence of the original strand of DNA? _____

_____

b. How could knowing the base sequence of a piece of DNA be important to a botanist? _____

_____

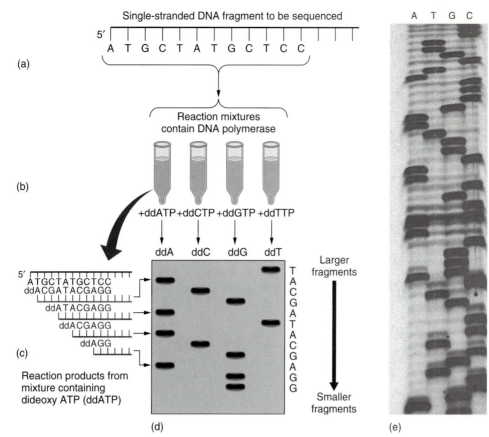

FIGURE 3.11

Determining the sequence of nucleotides in DNA. (*a*) A sample of the DNA to be sequenced is placed in each of four tubes. (*b*) The enzyme DNA polymerase is added to each tube along with a specific nucleotide-terminator. Polymerase replicates the DNA. The terminators are incorporated and will terminate various lengths of fragments of DNA. For example, the terminator ddATP (dideoxy adenosine triphosphate) will terminate a growing strand because it lacks a 3′ deoxy group. As the polymerase replicates the DNA, ddATP will halt the reaction wherever adenosine occurs. (*c*) Each tube will contain a sample of all possible replicated fragment lengths corresponding to the positions of that specific nucleotide. (*d*) During electrophoresis, the fragments migrate at different rates according to their length. (*e*) The lanes of the resulting gel are labeled according to their base: A, adenine; T, thymine; G, guanine; and C, cytosine.

Summary of Activities

1. Use column chromatography to separate a mixture of chemicals.

2. Use paper chromatography to separate a mixture of amino acids.

3. Use gel electrophoresis to separate a mixture of chemicals.

4. Interpret the pattern of chemical separation by a DNA-sequencing gel.

Questions for Further Thought and Study

1. How is gel electrophoresis similar to column chromatography? How is it different? What are the strengths and weaknesses of each technique?

2. What is the purpose of the buffer in gel electrophoresis? in column chromatography?

3. What kinds of molecules could biologists separate with the techniques you learned in this exercise?

Doing Botany Yourself

How would you modify the chromatography procedure to better resolve two amino acids that had approximately the same R$_f$ values?

Writing to Learn Botany

Which of the methods discussed in this exercise would best quantify the relative amounts of the molecules being separated? Why?

The Cell

Objectives

By the end of this exercise you will be able to

1. determine whether a cell is prokaryotic or eukaryotic on the basis of its structure
2. prepare a wet mount to view cells with a compound microscope
3. describe the structure and function of cellular organelles visible with a light microscope
4. stain and examine major cellular organelles.

Cells are the basic unit of living organisms because they perform all the processes we collectively call "life." All organisms are made of cells. Although most individual cells are visible only with the aid of a microscope, some may be a meter long (e.g., nerve cells) or as large as a small orange (e.g., the yolk of an ostrich egg). Despite these differences, all cells are designed similarly and share a variety of features. Before this lab, review in your textbook the general features of cellular structure and function and the functions of the major organelles.

PROKARYOTIC CELLS

Bacteria and cyanobacteria (fig. 4.1) are **prokaryotes** and thus do not contain a membrane-bounded nucleus or any other membrane-bounded **organelles.** Organelles are organized structures of macromolecules with a specialized function, and they are suspended in the **cytoplasm.** The cytoplasm of prokaryotes is enclosed in a **plasma membrane** (cellular membrane) and is surrounded by a supporting **cell wall** covered by a gelatinous **capsule. Flagella** and hairlike outgrowths called **pili** are common in prokaryotes. Within the cytoplasm of prokaryotes are **ribosomes** (small particles involved in protein synthesis), **mesosomes** (internal extensions of the plasma membrane), and **chromatin bodies** (concentrations of DNA). Prokaryotes do not reproduce sexually, but they have mechanisms for genetic recombination (see exercise 13).

Cyanobacteria

The largest prokaryotes are **cyanobacteria,** also called blue-green algae. They contain chlorophyll *a* and accessory pigments for photosynthesis, but these pigments are not contained in membrane-bound chloroplasts. Instead, the pigments are held in photosynthetic membranes called **thylakoids** (fig. 4.2). Cyanobacteria are often surrounded by a **mucilaginous sheath.**

Procedure 4.1
Examine cyanobacteria

1. Prepare a wet mount of *Oscillatoria*, a filament of cells, and one of *Gloeocapsa*, a loosely arranged colony. Review Procedure 1.5 in Exercise 1 for preparing a wet mount.
2. Focus with the low-power objective.
3. Rotate the high-power objective into place to see filaments and masses of cells.
4. Observe the cellular structures; draw the cellular shapes and relative sizes of *Gloeocapsa* and *Oscillatoria* in the space below.

   Gloeocapsa   *Oscillatoria*

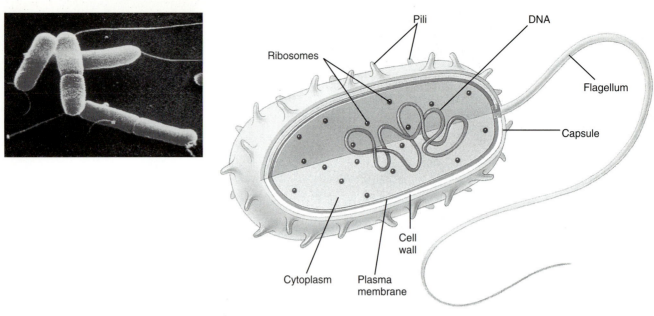

FIGURE 4.1

The structure of a bacterial cell. Bacteria lack a nuclear membrane. Not all prokaryotic (bacterial) cells have flagella but all do have ribosomes, plasma membrane, cytoplasm, and cell wall.

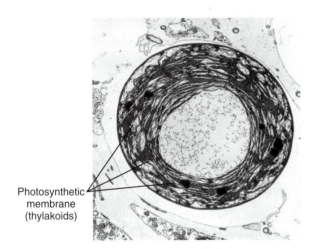

FIGURE 4.2

Electron micrograph of a photosynthetic bacterial cell, *Prochloron*. Extensive folded photosynthetic membranes are visible. The DNA is located in the clear area in the central region of the cell.

Question 1

a. Are the pigments in cyanobacterial cells localized or distributed uniformly? _____

b. Are nuclei visible in cyanobacterial cells? _____

c. Which of these two genera has the most prominent mucilaginous sheath? _____

d. How many cells are held within one sheath of *Gloeocapsa?* _____

Bacteria

Most bacteria are much smaller than cyanobacteria and do not contain chlorophyll. Examine some yogurt, which is a nutrient-rich culture of bacteria. The bacterial cells comprising most of the yogurt are *Lactobacillus,* a bacterium adapted to live on milk sugar (lactose). *Lactobacillus* converts milk to yogurt. Yogurt is acidic and keeps longer than milk. Historically, *Lactobacillus* has been used in many parts of the world by peoples deficient in lactase, an enzyme that breaks down lactose. Many Middle Eastern and African cultures use the more digestible yogurt in their diet instead of milk.

Procedure 4.2
Examine bacteria

1. Place a tiny dab of yogurt on a microscope slide.
2. Mix the yogurt in a drop of water, add a coverslip, and examine the yogurt with a compound microscope.
3. Focus with the low-power objective.
4. Rotate the high-power objective into place to see masses of rod-shaped cells.
5. Observe the simple external structure of bacteria and draw their cellular shapes in the space below.

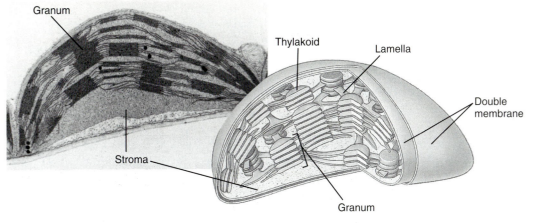

FIGURE 4.3

Chloroplast structure. The inner membrane of a chloroplast is fused to form stacks of closed vesicles called thylakoids. Photosynthesis occurs within these thylakoids. Thylakoids are typically stacked one on top of the other in columns called grana.

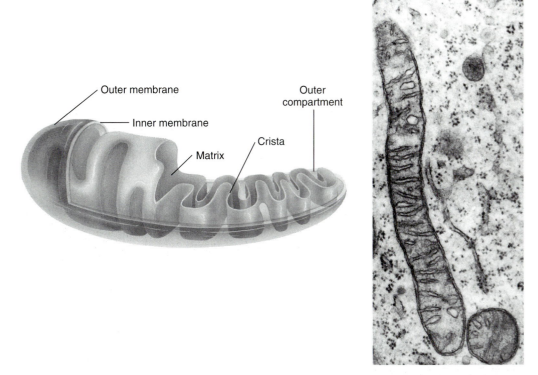

FIGURE 4.4

Mitochondria in longitudinal and cross section. These organelles are thought to have evolved from bacteria that long ago took up residence within the ancestors of present-day eukaryotes. Plant mitochondria tend to be shorter and thicker than animal mitochondria.

Question 2

How does the size of *Lactobacillus* compare with that of *Oscillatoria* and *Gloeocapsa*? _____

EUKARYOTIC CELLS

Eukaryotic cells contain membrane-bounded nuclei and other organelles. Nuclei contain genetic material of a cell and control cellular metabolism. **Cytoplasm** is the ground substance or matrix of the cell (see plate IV, fig. 7) and is contained by the plasma membrane. Within the cytoplasm are a variety of organelles. **Chloroplasts** are eliptical green organelles in plant cells (fig. 4.3). They are the site of photosynthesis; they are green because they contain chlorophyll, a photosynthetic pigment capable of capturing light energy. **Mitochondria** occur in plant and animal cells (fig. 4.4). These organelles are where aerobic respiration occurs. When

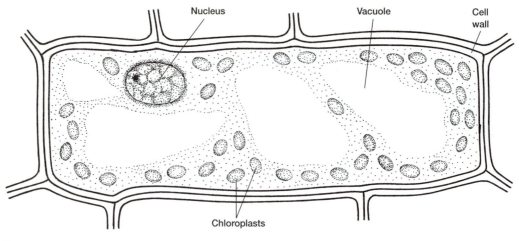

FIGURE 4.5

Elodea cell. The most prominent features of an *Elodea* cell are chloroplasti, vacuoler, the nucleus, and the cell wall.

viewed with a conventional light microscope, mitochondria are small, dark, and often difficult to see. All of the material and organelles contained by the plasma membrane is collectively called the **protoplast.**

Elodea Cells

Procedure 4.3
Examine living *Elodea* cells and chloroplasts

1. Remove a young leaf from the tip of a sprig of *Elodea*. *Elodea* is a common pond weed used frequently in studies of photosynthesis, cellular structure, and cytoplasmic streaming.

2. Place this leaf in a drop of water on a microscope slide with the top surface of the leaf facing up. The cells on the upper surface are larger and more easily examined. Add a coverslip. Be careful not to let the leaf dry. Add another drop of water if necessary.

3. Examine the leaf with your microscope. First use low then high magnification to bring the upper layer of cells into focus (fig. 4.5). Each of the small, regularly shaped units you see are cells surrounded by cell walls made of **cellulose** (see plate III, fig. 6). Cellulose is a complex carbohydrate made of glucose molecules attached end to end. The plasma membrane lies just inside the cell wall.

Question 3
a. What three-dimensional shape are *Elodea* cells? _____

_____

b. Examine various layers of cells by focusing up and down through the layers. About how many cells

thick is the leaf that you are observing? _____

_____

c. What are the functions of the cell wall? _____

_____

4. Chloroplasts appear as small green spheres within the cells. Locate cells with many chloroplasts, and estimate the number of chloroplasts in a healthy cell. Remember that because a cell is three-dimensional, some chloroplasts may obscure others.

Question 4
a. About how many chloroplasts are in an *Elodea* cell?

_____

b. What shape are the chloroplasts? _____

_____

c. How can you determine the three-dimensional shape of a chloroplast if the microscopic image is only two-

dimensional? _____

5. Determine the spatial distribution of chloroplasts within a cell. They may be pushed against the margins of the cell by the large **central vacuole,** containing mostly water and bounded by a **vacuolar membrane.** The vacuole occupies about 90% of the volume of a mature cell. Its many functions include storage of organic and inorganic molecules, ions, water, enzymes, and waste products.

6. Search for a **nucleus,** which may or may not be readily visible. Nuclei usually are appressed to the cell wall as a faint gray sphere. Staining the cells with a drop of iodine may enhance the nucleus. If your preparation is particularly good, a nucleolus may be visible as a dense spot in the nucleus.

7. Search for some cells that may appear pink due to water-soluble pigments called **anthocyanins.** These pigments give many flowers and fruits their bright reddish colors.

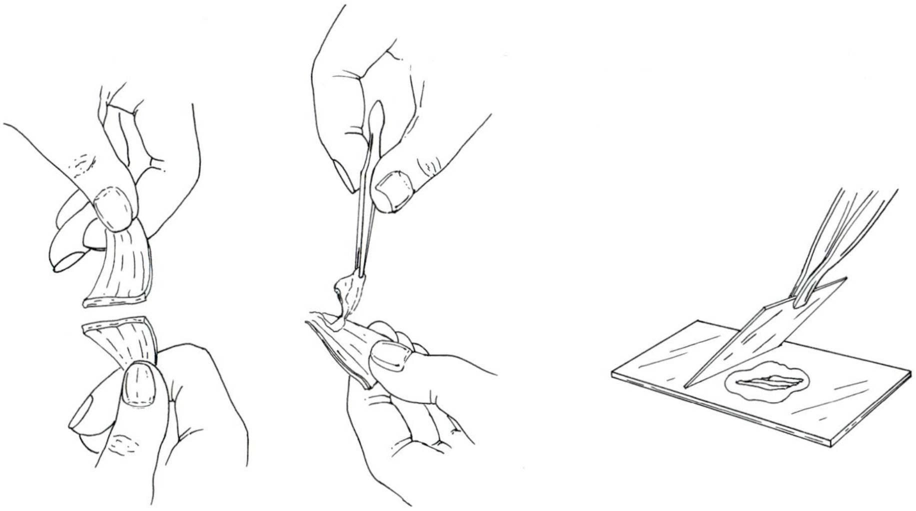

FIGURE 4.6

Preparing a wet mount of onion epidermis.

Question 5

a. Can you see nuclei in *Elodea* cells? _____

b. What is the function of nuclei? _____

_____

c. Which are larger, chloroplasts or nuclei? _____

_____

d. Why is the granular-appearing cytoplasm more apparent at the sides of a cell rather than in the middle? _____

_____

e. What is the function of the nucleolus? _____

_____

8. Search for movement of the chloroplasts. You may need to search many cells or make a new preparation. This movement is called **cytoplasmic streaming,** or **cyclosis.** Chloroplasts are not motile; instead, they are moved by the activity of the cytoplasm.

9. In the space below, sketch a few cells of *Elodea* and compare the cells with that shown in figure 4.5.

Question 6

a. Are all cellular components moving in the same direction and rate during cytoplasmic streaming?

_____

b. What do you conclude about the uniformity of

cytoplasmic streaming? _____

_____

Onion Cells

Staining often reveals the structure of cells more clearly. A specimen is **stained** by adding a dye, which preferentially colors some parts of the specimen but not others. Neutral red is a common stain that accumulates in the cytoplasm of the cell, leaving the cell walls clear. Nuclei appear as dense bodies in the translucent cytoplasm of the cells.

Procedure 4.4
Examine stained onion cells

1. Cut a red onion into fourths, and remove one of the fleshy leaves.

2. Snap the leaf backward, and remove the thin piece of epidermis formed at the break point (fig. 4.6) as demonstrated by your lab instructor.

3. Place this epidermal tissue in a drop of water on a microscope slide, add a coverslip, and examine the tissue.

4. Stain the onion cells by placing a small drop of 0.1% neutral-red at the edge of the coverslip and allowing it to diffuse over the onion.

5. Stain the tissue for 5 to 10 minutes.

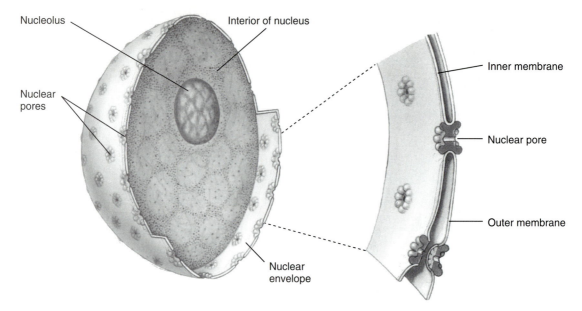

FIGURE 4.7

The nucleus. The nucleus consists of a double membrane, called a nuclear envelope, enclosing a fluid-filled interior containing the chromosomes. In cross section, the individual nuclear pores extend through the two membrane layers of the envelope; the dark material within the pore is protein, which controls access through the pore.

6. Carefully focus to distinguish the vacuole surrounded by the stained cytoplasm.
7. Search for the nucleus of a cell (fig. 4.7). The nucleus may appear circular in the central part of the cell. In other cells it may appear flattened.

_____

Question 7

How do you explain the differences in the apparent shapes and positions of the nuclei in different cells?

8. Repeat steps 1 to 7 and stain a new preparation of onion cells with other stains that are available.

_____

9. In the following space sketch a few of the stained onion cells.

Question 8

a. What shape are onion cells? _____
b. What cellular structures of onion are more easily seen in stained as compared to unstained

preparations? _____

c. Which of the available stains enhanced your

observations the most? _____

_____

d. Do onion cells have chloroplasts? _____

Why or why not? _____

_____

Mitochondria

Mitochondria are surrounded by two membranes (see plate V, fig. 9). The inner membrane folds inward to form **cristae,** which hold respiratory enzymes in place. Some DNA is also found in mitochondria. Chloroplasts (see following section on plastids) also are double-membraned and contain DNA.

Procedure 4.5
Examine mitochondria in onion cells

1. On a clean glass slide mix 2 to 3 drops of the stain Janus Green B with 1 drop of 7% sucrose.
2. Prepare a thin piece of onion epidermis (as instructed previously) and mount it in the staining solution. The preparation should be one cell thick. For mitochondria to stain well the onion cells must be healthy and metabolically active. Add a coverslip.
3. Search the periphery of cells to locate stained mitochondria, which are small blue spheres about 1 μm in diameter. The color will fade in 5 to 10 minutes, so make a new preparation if needed.

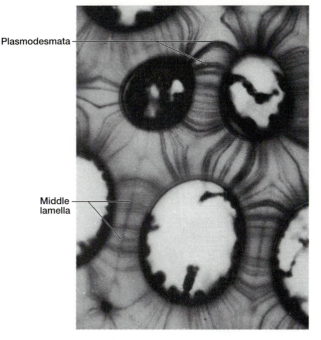

Plasmodesmata

Middle
lamella

FIGURE 4.8

The thickened primary cell walls of persimmon endosperm show distinctive plasmodesmata connecting adjacent cells. Middle lamella appear as faint lines parallel to the cell surface.

Cell Walls

Cell walls include an outer **primary cell wall** deposited during growth of the cell and a **middle lamella,** which is the substance holding walls from two adjacent cells together. The protoplasm of adjacent cells is connected by cytoplasmic strands called **plasmodesmata** that penetrate the cell walls (fig. 4.8).

Procedure 4.6
Examine cell walls and plasmodesmata

1. Prepare a wet mount of *Elodea* and examine the cell walls at high magnification. The middle lamella may be visible as a faint line between cells.

2. Obtain a prepared slide of persimmon (*Diospyros*) endosperm. The primary walls are highly thickened.

3. Locate the middle lamella as a faint line between cell walls.

4. Locate the plasmodesmata appearing as darkened lines perpendicular to the middle lamella and connecting the protoplasts of adjacent cells (fig. 4.8).

Plastids

Plastids are organelles where food is made and stored. You have already examined chloroplasts, a type of plastid in which photosynthesis occurs. Other plastids have different functions. We will examine two other kinds of plastids—amyloplasts and chromoplasts. **Amyloplasts** store starch

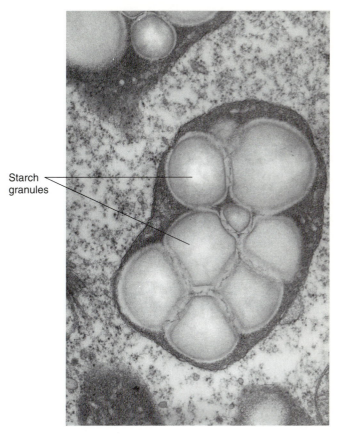

Starch
granules

FIGURE 4.9

Amyloplast with starch granules.

and stain darkly with iodine (fig. 4.9). **Chromoplasts** contain carotenoid pigments; they are easily observed as red plastids in red pepper. Many plant pigments are bound in chromoplasts.

Procedure 4.7
Examine amyloplasts

1. Use a razor blade to make a thin section of a potato tuber. Make the section as thin as you can.

2. Place the section in a drop of water on a microscope slide, and add a coverslip. Add another drop of water to the edge if needed.

3. Locate the small, clam-shaped amyloplasts within the cells. High magnification may reveal the eccentric lines distinguishing layers of deposited starch on the grains.

4. Stain the section by adding a drop of iodine to the edge of the coverslip. Iodine is a stain specific for starch (see exercise 2, "Biologically Important Molecules"). If necessary, pull the stain under the coverslip by touching a paper towel to the water at the opposite edge.

5. Examine the darkened amyloplasts.

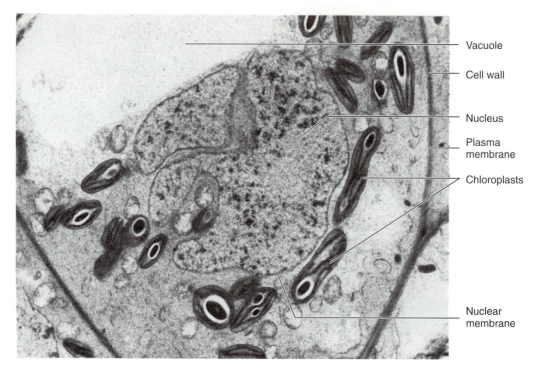

Vacuole

Cell wall

Nucleus

Plasma
membrane

Chloroplasts

Nuclear
membrane

FIGURE 4.10

Electron micrograph of a portion of a plant cell.

Question 9

a. Are any cellular structures other than amyloplasts
stained intensely by iodine? _____

_____

b. What can you conclude about the location of starch
in storage cells of potato? _____

_____

c. Why are potatoes a good source of carbohydrates?

_____

Procedure 4.8
Examine chromoplasts

1. Use a razor blade to prepare a thin section of a red
pepper.
2. Place the section in a drop of water on a microscope
slide, add a coverslip, and focus under low, then
higher magnification.
3. Examine the cellular structure and locate
chromoplasts.
4. If carrots or yellow or orange flower petals are
available, make thin sections and search for
chromoplasts. Determine the shape, size, and color of
the chromoplasts.

Question 10

What are the functions of chromoplasts? _____

_____

CELLULAR ULTRASTRUCTURE

As mentioned in the lab exercise on microscopy (exercise 1),
not all cellular structures are visible with a light microscope
due to its limited resolution. Electron microscopes improve
resolution approximately 1,000-fold and reveal cellular structures not easily seen with a light microscope.

Examine the electron micrograph of plant cells
shown in figure 4.10. Refer to your textbook to describe
the function of cellular structures visible in this micrograph. In your textbook locate and examine the micrographs presented in the chapter on cellular structure.

Summary of Activities

1. Examine cyanobacteria.
2. Examine bacteria in yogurt.
3. Examine *Elodea* cells.
4. Prepare and examine stained onion cells.
5. Prepare and examine stained mitochondria in onion
cells.

6. Examine amyloplasts and chromoplasts.
7. Examine electron micrographs of cellular ultrastructure.

Questions for Further Thought and Study

1. What is a cell?
2. Describe the structure and function of each component of the cells that you observed with a light microscope.
3. What is the function of cytoplasmic streaming?
4. What are the advantages and disadvantages to an individual cell of being part of a multicellular organism?

Doing Botany Yourself

Determine the total surface area of the chloroplasts in a typical *Elodea* cell. Assume that each chloroplast is a sphere of 5 = μm diameter (the surface area of a sphere = πd^2). What is the significance of this surface area? Would it be advantageous for a cell to be filled with chloroplasts? Why or why not?

Writing to Learn Botany

What criteria might you use to distinguish colonial organisms such as many cyanobacteria from truly multicellular organisms?

Cellular Membranes

5

Objectives

By the end of this exercise you will be able to

1. discuss the function and structure of cellular membranes
2. discuss the relationship of structure and function in membranes
3. predict the effect of common organic solvents and extreme temperatures on membrane integrity
4. relate the results of experiments with beet membranes to the general biology of membranes.

Membranes separate and organize the myriad reactions within cells and allow communication with the surrounding environment. Although membranes are only a few molecules thick (5–10 nm), they (1) retard diffusion of certain molecules, (2) house receptor molecules that detect other cells or organelles, (3) provide sites for active and passive transport of selected molecules, and (4) organize life processes.

As with all other biological entities, the structure of a membrane reflects its function. Membranes are made of a **phospholipid bilayer** of molecules interspersed with protein molecules. A phospholipid molecule consists of a phosphate group and two fatty acids bonded to a three-carbon glycerol chain (fig. 5.1). Phospholipids are also polarized, and polar molecules are charged with positive and negative areas. In phospholipids the phosphate group is polar and **hydrophilic** ("water-loving"), whereas the fatty-acid chains are nonpolar and **hydrophobic** ("water-fearing"). These polarized phospholipids have a natural tendency to self-assemble into a double-layered sheet. In this double layer, the hydrophobic tails of lipids form the core of the membrane, and the hydrophilic phosphate groups line both surfaces (figs. 5.2 and 5.3). This elegant assembly is stable and self-repairing and resists penetration by most hydrophilic molecules.

Membrane structure also includes proteins that are dispersed as a mosaic throughout the fluid bilayer of lipids (fig. 5.3). These membrane proteins are not fixed in position; they move about freely and may be densely packed in some membranes and sparsely distributed in others. Carbohydrate chains (strings of sugar molecules) are often bound to these proteins and to lipids. These chains serve as distinctive identification tags, unique to particular types of cells. This elaborate combination forms the **fluid mosaic model** of membrane structure. Membranes are not entirely impenetrable. Proteins embedded in the phospholipid bilayer can selectively take up and expel molecules that otherwise could not penetrate the membrane. In doing so, these proteins function as pores, permitting and often facilitating passage of specific ions and polar molecules. In addition to forming pores and sites for active transport, membrane-bounded proteins also function as enzymes and receptors that detect signals from the environment or from other cells.

The physical and chemical integrity of a membrane is crucial for the proper functioning of the cell or organelle that it surrounds. As a stable sheet of interlocking molecules, the membrane functions as a barrier to simple diffusion (fig. 5.4). Membrane permeability depends on the solubility and size of the molecules trying to penetrate the membrane. Molecules that are lipid-soluble pass easily through the lipid core of a membrane.

Question 1

What ions must routinely move across cell

membranes? _____

_____

In this exercise you will physically and chemically stress the integrity of membranes.

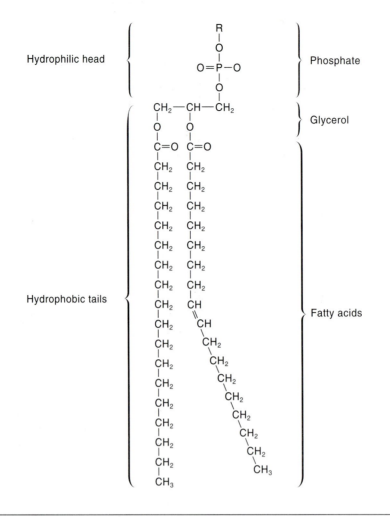

FIGURE 5.1

The structure of a phospholipid. A phospholipid consists of glycerol that is bonded to two fatty acids, which are hydrophobic, and to one phosphate group, which is hydrophilic. Phospholipids vary by their fatty acids and by the side chains (R groups) that are attached to the phosphate. The R groups include glycerol, sugars, and amine-containing carbon chains.

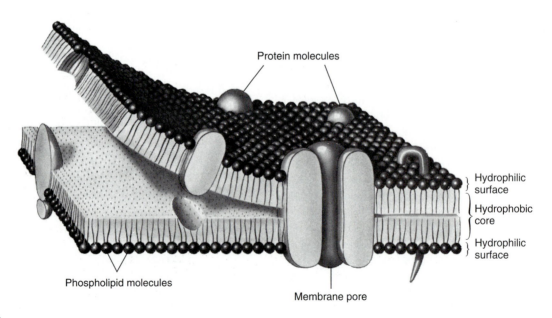

FIGURE 5.2

Model of a bilayer membrane.

FIGURE 5.3

Electron micrograph of a bilayered membrane surrounding a cell. Arrows indicate each layer.

Large, uncharged polar molecules	Glucose	
Ions	H^+, Na^+, HCO_3^-, K^+, Ca^{2+}, Cl^-, Mg^{2+}	
Hydrophobic molecules	Hydrocarbons, O_2	
Small, uncharged polar molecules	H_2O, CO_2	

FIGURE 5.4

Phospholipid bilayers are selectively permeable. The bilayer is much more permeable to hydrophobic molecules and small uncharged polar molecules than it is to ions and large polar molecules.

BEET CELLS AS AN EXPERIMENTAL SYSTEM

Beet tissue will be your model to investigate membrane integrity. Roots of beet (*Beta vulgaris*) contain large amounts of a reddish pigment called **betacyanin** localized almost entirely in the large central vacuoles of cells. Betacyanin in healthy cells remains inside the vacuoles, which are surrounded by a vacuolar membrane called the **tonoplast.** The entire cell (including vacuole, tonoplast, and cytoplasm) is surrounded by a plasma membrane.

In the following two experiments you will subject beet cells to a range of temperatures and organic solvents and determine which treatments stress and damage the membranes the most. If stress damages the membranes, betacyanin will leak through the tonoplast and plasma membrane. This leakage will produce a red color in the water surrounding the stressed beet. Thus, you can monitor membrane damage by monitoring the intensity of color resulting from a particular treatment.

THE EFFECT OF TEMPERATURE STRESS ON MEMBRANES

Extreme temperatures destroy a membrane as a physical barrier to diffusion. High temperatures cause such violent molecular collisions that the assembled membrane is torn apart. Freezing temperatures cause water to crystallize as ice and expand due to the hydrogen bond alignment. This expansion and ice formation can rupture membranes.

Procedure 5.1
Observe the effect of temperature stress on cellular membranes

1. Cut six uniform cylinders of beet using a cork borer with a 5 mm inside diameter. Trim each cylinder to 15 mm in length.

2. Place these cylinders of beet tissue in a beaker and rinse them with tap water for 2 minutes to wash betacyanin from the injured cells on the surface. Discard the colored rinse water.

TABLE 5.1

COLOR INTENSITY OF BETACYANIN LEAKED FROM DAMAGED MEMBRANES TREATED
AT SIX TEMPERATURES

Tube	Treatment (°C)	Color Intensity (0–10)	Absorbance (460 nm)
1	70		
2	55		
3	40		
4	20		
5	5		
6	–5		

3. Place one of the six beet sections into each of six dry test tubes.

4. Label the tubes 1 to 6 and write the temperature treatment on each tube as listed in table 5.1.

5. FOR COLD TREATMENTS:

a. Place tube 5 in a refrigerator (5°C) and tube 6 in a freezer (–5°C). If a refrigerator and freezer are not available in your lab, then give your labeled tubes to an assistant who will take them to another facility.

b. Leave tubes 5 and 6 in the cold for 30 minutes. While waiting, proceed with hot treatments (step 6). However, watch your time and return to steps 5c and 5d after 30 minutes.

c. After 30 minutes, remove the beets from the freezer and refrigerator and add 10 mL of distilled water at room temperature to each of the tubes.

d. Let the cold-treated beets soak in distilled water for 20 minutes. Then remove and discard the beets from tubes 5 and 6.

6. FOR HOT TREATMENTS:

a. Take the beet section out of tube 1 and immerse it in a beaker of hot water at 70°C for 1 minute. Hot tap water should be adequate, but carefully adjust the temperature to 70°C. Handle the beet with forceps, but don't squeeze it tightly because you may rupture cells.

b. After 1 minute at 70°C, return the beet to tube 1 and add 10 mL of distilled water at room temperature.

c. Cool the beaker of hot water to 55°C and immerse the beet from tube 2 for 1 minute. Return the beet to tube 2 and add 10 mL of distilled water at room temperature.

d. Cool the beaker of hot water to 40°C and immerse the beet from tube 3 for 1 minute. Return the beet to tube 3 and add 10 mL of distilled water at room temperature.

e. Cool the beaker of hot water to 20°C and immerse the beet from tube 4 for 1 minute. Return the beet to tube 4 and add 10 mL of distilled water at room temperature.

f. Allow the treated beets in tubes 1 to 4 to soak in distilled water at room temperature for 20 minutes. Then remove and discard the beets and measure the extent of membrane injury according to the amount of betacyanin that diffused into the water.

7. FOR ALL SIX TEMPERATURE TREATMENTS: Quantify the relative color of each solution between 0 (colorless) and 10 (darkest red), and record your results in table 5.1. If color standards are available, examine them to determine relative values for the colors of your samples.

Use the graph paper at the end of this exercise to graph *Temperature* versus *Relative Color* according to a demonstration graph provided by your instructor. Your instructor may also ask you to quantify your results further using a spectrophotometer. If so, see appendix B for instructions for using a spectrophotometer. Read the absorbance of the solutions at 460 nm and record your results in table 5.1. Then graph Temperature versus Absorbance.

Question 2

a. Which temperature damaged the membranes the

most? _____

the least? _____

b. In general, which is more damaging to membranes,

extreme heat or extreme cold? _____

Why? _____

c. If the results of this experiment are easily observed with the unaided eye, why use a spectrophotometer?

_____

_____

d. The beets were subjected to cold temperatures longer than to hot temperatures to make sure that the beet sections were thoroughly treated. Why does the

freezing treatment require more time? _____

_____

TABLE 5.2

COLOR INTENSITY OF BETACYANIN LEAKED FROM DAMAGED CELLS TREATED WITH VARIOUS CONCENTRATIONS OF TWO ORGANIC SOLVENTS

Tube	Treatment	Color Intensity 0–10	Absorbance (460 nm)
1	1% acetone		
2	25% acetone		
3	50% acetone		
4	1% methanol		
5	25% methanol		
6	50% methanol		

THE EFFECT OF ORGANIC SOLVENT STRESS ON MEMBRANES

Organic solvents will dissolve a membrane's lipids. Acetone and alcohol are common solvents for various organic molecules, but acetone has the greater ability to dissolve lipids.

CAUTION

Organic solvents are flammable. Extinguish all open flames and heating elements.

Question 3

Which of the two organic liquids (acetone or methanol) do you predict will damage membranes the most? _____

Why? _____

Procedure 5.2

Observe the effect of organic solvents on cellular membranes

1. Cut six uniform cylinders of beet using a cork borer with a 5 mm inside diameter. Trim each cylinder to 15 mm in length.
2. Place these cylinders of beet tissue in a beaker and rinse them with tap water for 2 minutes to wash betacyanin from the injured cells on the surface. Discard the colored rinse water.
3. Place one of the six beet sections into each of six dry test tubes.
4. Label the tubes 1 to 6 and write the organic solvent treatment on each tube as listed in table 5.2.
5. Add 10 mL of the appropriate solvent (see table 5.2) to each of the six tubes.

6. Keep all beets at room temperature for 20 minutes and shake them occasionally. After 20 minutes, remove and discard the beet sections and measure the extent of membrane damage according to the amount of betacyanin that diffused into the water.
7. Quantify the relative color of each solution between 0 (colorless) and 10 (darkest red). Record your data in table 5.2. If color standards are available, examine them to determine the relative values for the colors of your samples.

Graph *Concentration of Organic Solvent* versus *Relative Color* according to a demonstration graph provided by your instructor. Your instructor may also ask you to further quantify your results using a spectrophotometer. If so, see appendix B for instructions for using a spectrophotometer. Read the absorbance of the solutions at 460 nm and record your results in table 5.2. Then graph the *Concentration of Organic Solvent* versus *Absorbance*.

Question 4

a. Based on your results, are lipids soluble in both acetone and methanol? _____

b. Based on your results, which damages membranes more, 50% methanol or 25% acetone? _____ _____

c. Which solvent has the greatest lipid solubility? _____ _____

d. The concentration of solvent affects its ability to dissolve lipids. Based on your results, did the highest concentration of both solvents cause the most damage? _____

e. What other solvents might be interesting to test in this experiment? _____ _____

Summary of Activities

1. Observe damage to temperature-stressed membranes.
2. Graph membrane treatment (temperature) versus damage (absorbance).
3. Observe damage to membranes stressed by organic solvents.
4. Graph membrane treatment (solvent concentration) versus damage (absorbance).

Questions for Further Thought and Study

1. Are your conclusions about membrane structure and stress valid only for beet cells? Why or why not?
2. What characteristics of beets make them useful as experimental models for studying cellular membranes?

3. Freezing temperatures are often used to preserve food. Considering the results of this experiment, which qualities of food are preserved and which are not?
4. Movement of water through membranes has long puzzled scientists. Why would you not expect water to move easily through a membrane?

Doing Botany Yourself

How would you design an experiment to determine the relative lipid solubilities of various organic solvents?

Writing to Learn Botany

What role did the stability and tendency for self-assembly play in the early evolution of life?

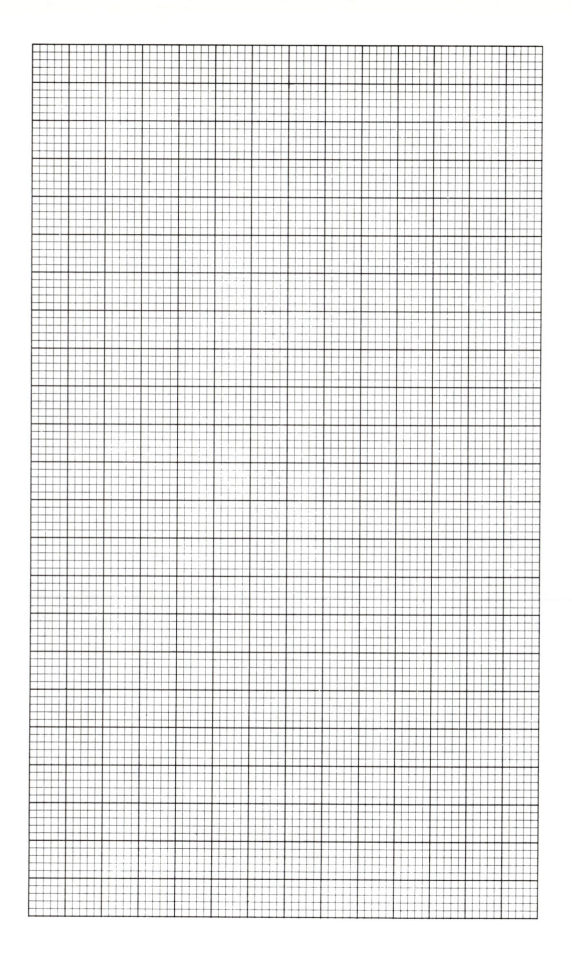

Diffusion and Osmosis

Objectives

By the end of this exercise you will be able to

1. describe the motion of molecules as demonstrated by Brownian movement
2. list the factors controlling a substance's diffusion and contributing to its free energy
3. predict the direction and relative rates of diffusion of molecules of different sizes and concentrations
4. describe how a semipermeable membrane affects the diffusion of molecules
5. predict the direction and rate of osmosis in and out of cells in hypotonic, hypertonic, and isotonic environments
6. describe how a hypertonic solution affects the volume and integrity plant cells.

All molecules display random thermal motion, or kinetic energy, and a dissolved molecule tends to move around in a solution. As a result, molecules diffuse outward from regions of high concentration to regions of lower concentrations. This random movement continues until the distribution of molecules becomes homogenous throughout the solution.

For example, when a crystal of dye is placed in a container of water, the dye disperses as the crystal dissolves. The rate of dispersal depends on the size of the dye molecules and the temperature of the solution. Regardless of this rate, the dye eventually becomes uniformly distributed throughout the solution. This phenomenon is easily illustrated by placing a drop of dye into a glass of water. Because biological systems are mostly water, the movement of materials within solutions and across barriers is critical to the typical functions of cells. In this exercise you will experiment with the diffusion of molecules in artificial and living systems.

Heat causes **random motion** of molecules and passively moves molecules in biological systems. Although we cannot directly see molecules move, we can see small particles move after they collide with moving molecules. This motion of particles is called **Brownian movement** and is visible using your microscope's high magnification. India ink is a convenient source of small particles for demonstrating Brownian movement. It is not a solution, as you might suspect; rather, it is a suspension of small pieces of black pigment. These pieces are small enough to vibrate when water molecules bump into them.

Procedure 6.1
Observe Brownian movement

1. Place a small drop of dilute India ink on a microscope slide and cover the drop with a coverslip.
2. Focus first at low magnification, and then rotate to higher power (×40). Be careful not to get ink on the objective lens.
3. Fine-focus the image. At first the field of view appears uniformly gray. With sharp focus, however, you will see thousands of small particles vibrating rapidly.
4. Leave the microscope light on. Observe any changes in motion with increased heat.

Question 1
a. Briefly describe your observation of the moving ink particles. _____
b. Is the motion of water molecules or India ink molecules responsible for most of the energy for Brownian movement? _____
c. Does particle movement change visibly with heat?
_____
How? _____

DIFFUSION

In biological systems, substances often move through solutions and across membranes in a predictable direction. This passive, directional movement of molecules is **diffusion.** The *direction* of diffusion depends on gradients of concentration, heat, and pressure. Specifically, molecules diffuse from an area of high concentration, heat, and pressure to an area of low concentration, heat, and pressure. The *rate* of diffusion is determined by the steepness of the gradient and other factors such as molecular weight.

Concentration, heat, and pressure contribute to the overall free energy of a substance. **Free energy** is the energy available in a system to do work; substances diffuse from an area of high free energy to an area of low free energy. Therefore, *diffusion is best defined as the net movement of molecules from an area of high free energy to an area of low free energy.*

Even though temperature, pressure, and concentration all affect diffusion, temperature and pressure are relatively constant in most biological systems. Therefore, concentration is usually the best predictor of a substance's free energy and, therefore, direction of diffusion. But remember that temperature and pressure gradients may also affect diffusion.

Diffusion and Molecular Weight

Twenty-four hours before your class meeting your instructor inoculated some petri plates containing agar with potassium permanganate (molecular weight = 158 g mole^{-1}) and malachite green (molecular weight = 929 g mole^{-1}).

Procedure 6.2
Observe diffusion as affected by molecular weight

1. Examine one of the prepared agar plates and note the two halos of color. These halos indicate that the chemicals have diffused away from the two original spots and moved through the agar.
2. Measure the halos with a ruler.
3. On the following circle draw and label the halos for the two chemicals as they appear in the circular petri plate.

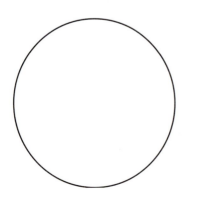

Question 2

a. Would a substance of high or low molecular weight diffuse faster? _____

   Why? _____

b. Considering the different molecular weights of potassium permanganate and malachite green, which should have the larger halo after the same amount of time? _____

   Why? _____

c. Do molecules stop moving when diffusion stops?

   _____

d. Would a higher temperature change the diffusion rates and the results of the experiment? _____

   _____

DIFFUSION AND DIFFERENTIALLY PERMEABLE MEMBRANES

Membranes surround plant cells and organelles and organize an immense number of simultaneous reactions. However, the barrier imposed by a cellular membrane does not completely isolate the cell. Instead, it allows a cell to selectively communicate with its environment. Membranes are "alive" in the sense that they respond to their environment and allow some molecules to pass while retarding others. Thus, membranes are selective and **differentially permeable** (fig. 6.1). This selective permeability results from the basic structure of membranes. They have a two-layered core of nonpolar lipid molecules. This core selects against molecules that are not readily soluble in lipids.

Two important characteristics of molecules govern their passive movement through a lipid membrane: polarity and size. **Polar molecules** exhibit charge separation (i.e., they have positively charged areas and negatively charged areas). They do not pass through the membrane's nonpolar, lipid core easily unless the polar molecules are small (e.g., water). **Nonpolar molecules** have no local areas of positive or negative charge. Small nonpolar molecules pass through membranes easily, while large polar molecules are retained.

In general, small molecules pass through a membrane more easily than do large molecules. We can demonstrate membrane selection for molecular size by using a bag made from **dialysis tubing** to model a differentially permeable membrane surrounding a plant cell. **Dialysis** is the separation of dissolved substances by means of their unequal diffusion through a differentially permeable membrane. Dialysis membranes (or tubing) are good models of differentially permeable membranes because they have small pores that allow small molecules such as water molecules to pass but block large molecules such as glucose.

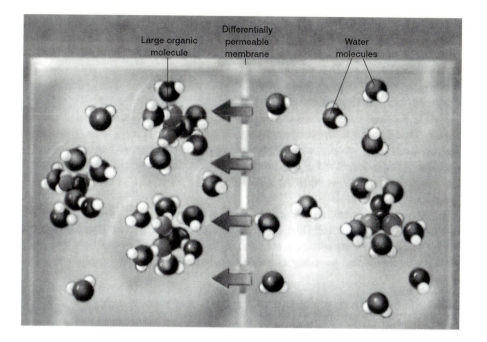

FIGURE 6.1

A differentially permeable membrane prevents the movement of some molecules but not others. Arrows indicate water moving from an area of high concentration (high free energy) to an area of lower concentration (lower free energy).

Examine some dialysis tubing. Although the dried material looks like a narrow sheet of cellophane it is actually a flattened, open-ended tube.

In procedure 6.3 you will use two indicators: **phenolphthalein** and **iodine.** Phenolphthalein is a pH indicator that turns red in basic solutions (see plate II, fig. 3). Iodine is a starch indicator that changes from light brown to dark blue in the presence of starch (see Exercise 2, "Biologically Important Molecules").

Procedure 6.3
Observe diffusion across a selective membrane

1. Obtain two pieces of water-soaked dialysis tubing approximately 15 cm long and four pieces of string.

2. Seal one end of each bag by folding over 1 to 2 cm of the end and tying it tightly with a piece of string. The ends of the tube must be sealed tightly to prevent leaks (fig. 6.2).

3. Roll the untied end of each tube between your thumb and finger to open it and form a bag.

4. Fill one tube with 10 mL of water and add 3 drops of phenolphthalein. Seal the open end of the bag by folding the end and tying it securely.

5. Fill the other bag with 10 mL of starch suspension. Seal the open end of the bag by folding the end and tying it securely.

6. Gently rinse the outside of each bag in tap water.

FIGURE 6.2

A folded and tied dialysis bag used as a cellular model.

7. Fill a beaker with 200 mL of tap water and add 10 drops of 1 M sodium hydroxide. Submerge the dialysis bag containing phenolphthalein in the beaker.

8. Fill a beaker with 200 mL of tap water and add 20 to 40 drops of iodine. Submerge the dialysis bag containing starch in the beaker.

9. Observe color changes in the two bags' contents and the surrounding solutions.

10. Record in figure 6.3 the color inside and outside the bags. Label the contents inside and outside the bags.

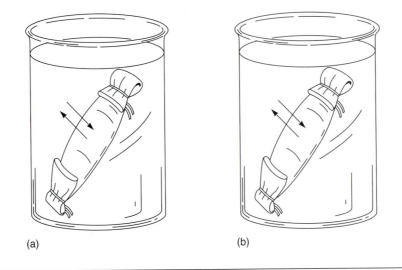

(a)   (b)

FIGURE 6.3

(a) Movements and reaction of sodium hydroxide and phenolphthalein through a differentially permeable membrane. (b) Movements and reaction of iodine and starch. Record the results of your experiment on this diagram.

Question 3

a. Describe color changes in the two bags and their

surrounding solutions. _____

_____

b. For which molecules and ions (i.e., phenolphthalein, iodine, starch, Na⁺, or OH⁻) does your experiment give evidence for passage through the semipermeable

membrane? _____

_____

c. What characteristic distinguishes those molecules and ions that pass through the membrane from those

that do not pass through the membrane? _____

_____

d. What determined the direction of net movement of the ions—the nature of the membrane, the

concentration of the ions, or both? _____

_____

OSMOSIS AND THE RATE OF DIFFUSION ALONG A FREE-ENERGY GRADIENT

The speed at which a substance diffuses from one area to another depends on the free-energy gradient between those areas. For example, if concentrations of a diffusing substance at the two areas differ greatly, then the free-energy gradient is steep and diffusion is rapid. Conversely, when the free energy of a substance at the two areas is equal, then the diffusion rate is zero and net movement of the substance stops.

Osmosis is diffusion of water across a differentially permeable membrane in living organisms. Osmosis follows the same laws as diffusion but always refers to water, the principal solvent in cells. A **solution** is a homogenous liquid mixture of two or more kinds of molecules. A **solvent** is a fluid that dissolves substances, and a **solute** is a substance dissolved in a solution. Because water is the universal solvent in cells, its diffusion across differentially permeable membranes is critical to living organisms.

We can simulate osmosis by using dialysis bags to model cells under different conditions and measuring the direction and rate of osmosis. Each of the four dialysis bags in the following experiment is a model of a cell. Bag A simulates a cell with a solute concentration that is hypotonic relative to its environment. **Hypotonic** describes a solution with a lower concentration of solutes. Water moves across semipermeable membranes out of hypotonic solutions. Conversely, the solution surrounding bag A is hypertonic relative to the cell. **Hypertonic** refers to a solution with a higher concentration of solutes.

Bag B represents a cell whose solute concentration equals the concentration in the environment; that is, this cell (bag B) is isotonic to its environment. **Isotonic** refers to two solutions that have equal concentrations of solutes. Bags C and D are both hypertonic to their environment and have higher solute concentrations than the surrounding environment.

Procedure 6.4
Observe osmosis along a free-energy gradient

1. Obtain four pieces of water-soaked dialysis tubing 15 cm long and eight pieces of string. Seal one end of each tube by folding and tying it tightly.

2. Open the other end of the tube by rolling it between your thumb and finger.

6-4

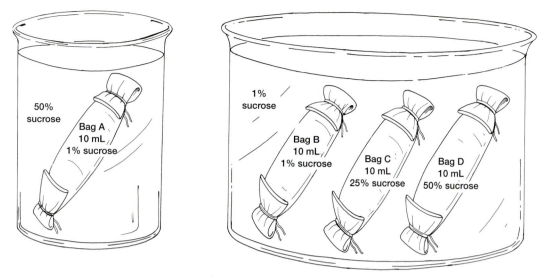

FIGURE 6.4

Experimental setup for four cellular models used to measure the rate of osmosis.

TABLE 6.1

CHANGES IN WEIGHT OF DIALYSIS BAGS USED AS CELLULAR MODELS

| | 0 Min | 15 Min | | 30 Min | | 45 Min | | 60 Min | |
	Initial Wt.	Total Wt.	Change in Wt.	Total Wt.	Change in Wt.	Total Wt.	Change in Wt.	Total Wt.	Change in Wt.
Bag A									
Bag B									
Bag C									
Bag D									

Note: Each change in weight is for the previous 15-minute interval.

3. Fill the bags with the contents shown in figure 6.4. To label each bag, insert a small piece of paper with the appropriate letter (A, B, C, or D) written on it in pencil.

4. For each bag, loosely fold the open end and press on the sides to push the fluid up slightly and remove most of the air bubbles. Tie the folded ends securely, rinse the bags, and check for leaks.

5. Blot excess water from the outside of the bags and weigh each bag to the nearest 0.1 g.

6. Record these initial weights in table 6.1 in the first column.

7. Place bags B, C, and D in a beaker or large bowl filled with 1% sucrose (fig. 6.4). Record the time.

8. Place bag A in an empty beaker and fill the beaker with just enough 50% sucrose to cover the bag. Record the time.

9. Remove the bags from the beakers at 15-minute intervals for the next hour, blot them dry, and weigh

them to the nearest 0.1 g. Handle the bags delicately to avoid leaks, and quickly return the bags to their respective containers.

10. For each 15-minute interval record the total weight of each bag and its contents in table 6.1. Then calculate and record in table 6.1 the change in weight since the previous weighing.

Procedure 6.5
Graph diffusion

1. Construct a graph with *Total Weight (g)* on the vertical axis and *Time (min)* on the horizontal axis. Use the graph paper at the end of this exercise.

2. Plot the data for total weight at each time interval from table 6.1.

3. Include the data for all four bags as four separate curves on the same graph.

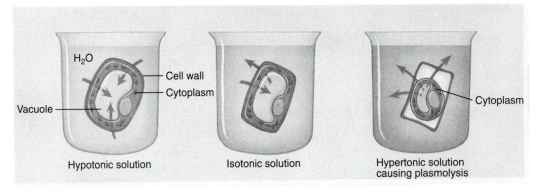

FIGURE 6.5

Plasmolysis occurs when water moves from the cell into a surrounding hypertonic solution. This causes the cytoplasm to shrink and pull in from the cell wall. Arrows show the flow of water.

Question 4

a. Did water move across the membrane in all bags containing solutions of sugar? _____

b. In which bags did osmosis occur? _____

c. A free-energy gradient for water must be present in cells for osmosis to occur. Which bag(s) represented the steepest free-energy gradient for water relative to its surrounding environment? _____
_____

d. The steepest gradient of free energy should result in the highest rate of diffusion. Examine the data in table 6.1 for the 15- and 30-minute intervals. Did the greatest change in weight occur in cells with the steepest free-energy (concentration) gradient?
_____
Why or why not? _____
_____

e. If the solution surrounding bags B, C, and D was 50% sucrose instead of 1%, which bag would exhibit the highest rate of diffusion? _____

f. Refer to your graph. How does the slope of a segment of a curve relate to the rate of diffusion? _____
_____

g. What component of free energy causes the curves for bags C and D eventually to become horizontal (i.e., slope = 0)? _____

PLASMOLYSIS OF PLANT CELLS

Plasmolysis is the shrinking of the cytoplasm of a plant cell in response to diffusion of water out of the cell and into a hypertonic solution (high-salt concentration) surrounding the cell (fig. 6.5). During plasmolysis the cellular membrane pulls away from the cell wall. In the next experiment you will examine the effects of highly concentrated solutions on diffusion and cellular contents.

Procedure 6.6
Observe plasmolysis

1. Prepare a wet mount of a thin layer of onion epidermis or *Elodea* leaf.
2. Add 2 to 3 drops of 40% NaCl to one edge of the coverslip.
3. Wick this salt solution under the coverslip by touching a piece of absorbent paper towel to the fluid at the opposite edge of the coverslip.
4. Examine the cells and note that the cytoplasm is no longer pressed against the cell wall. This shrinkage is plasmolysis.

Question 5

a. Why did the plant cells plasmolyze when immersed in a hypertonic solution? _____
_____

b. What can you conclude about the permeability of the cell membrane (the membrane surrounding the cytoplasm) and vacuolar membrane (the membrane surrounding the vacuole) to water? _____
_____

To observe the effects of cellular plasmolysis on a larger scale, compare petioles of celery that have been immersed in distilled water or in a salt solution overnight.

Question 6

What causes crispness in celery? _____
_____

Summary of Activities

1. Observe Brownian movement.
2. Observe how molecular weight affects the rate of diffusion.
3. Observe diffusion across a differentially permeable membrane.
4. Observe osmosis along a free-energy gradient.
5. Graph rate of diffusion.
6. Observe plasmolysis of plant cells.

Questions for Further Thought and Study

1. Why must particles be extremely small to demonstrate Brownian movement?
2. What is the difference between molecular motion and diffusion?

3. Your data for diffusion of sucrose across a differentially permeable membrane could be graphed with *Change in Weight* on the vertical axis rather than *Total Weight*. How would you interpret the slope of the curves produced?
4. How do cells such as aquatic algae and protozoans avoid lysis in freshwater? (Lysis is the rupture of a cell due to inflow of water.)
5. If you immerse your hand in distilled water for 15 minutes, will your cells lyse? Why or why not?

Doing Botany Yourself

Consider that water moves in and out of cells along a concentration gradient. How would you design an experiment to determine the solute concentration within cells of a piece of apple?

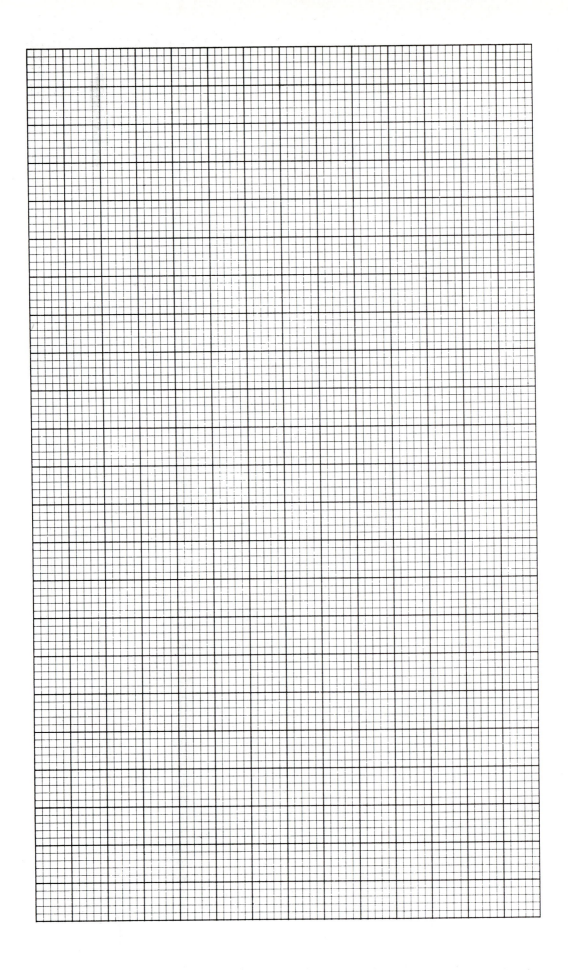

Enzymes

Objectives

By the end of this exercise you will be able to

1. understand the concept of optimal conditions and how it relates to enzymatic activity
2. predict how changes in temperature, pH, substrate concentration, and enzyme concentration affect the time to complete enzymatic reactions
3. explain the mechanism by which temperature, pH, substrate concentration, and enzyme concentration alter enzymatic reactions.

Fortunately, all chemical reactions within our cells do not occur spontaneously. If they did, our metabolism would be chaotic. Instead, most molecules are relatively unreactive, and metabolic reactions must be catalyzed by proteins called **enzymes.** Enzymes are biocatalysts, meaning that they accelerate reaction rates to biologically useful rates. Enzymes act by binding to reacting molecules, called the **substrate,** to form an **enzyme-substrate** complex. This complex stresses or distorts chemical bonds to form a **transition state** in which the substrate becomes more reactive and the metabolic reaction continues. The energy necessary to form the transition state is called **energy of activation** and is supplied by the enzyme. The site of attachment and the surrounding parts of the enzyme that stress the substrates' bonds constitute the enzyme's **active site.**

The reaction is complete when the **product** forms and the enzyme is released in its original condition.

The enzyme then repeats the process with other molecules of substrate (fig. 7.1).

Enzymes are proteins made of long chains of amino acids that form complex shapes. Although cells contain many enzymes, each kind of enzyme has a precise structure and function, and each enzyme catalyzes a specific reaction. This specificity results from an enzyme's unique structure and shape. For example, the shape of the active site on the enzyme's surface is unique and usually couples with only one kind of substrate. Any structural change may **denature** or destroy the enzyme's effectiveness by altering the active site and slowing down the reaction rate. For this reason, the rate of an enzymatic reaction depends on conditions in the immediate environment. These conditions affect the shape of the enzyme and modify the active site and precise fit of an enzyme and its substrate.

In this exercise you will learn that environmental factors such as temperature and pH affect enzymatic reactions. The range of values at which an enzyme functions best represent that enzyme's **optimal conditions.** Other factors such as the amount of substrate or concentration of enzyme also affect the reaction rate. The optimal conditions for the enzymes of an organism may be specific for that species and usually are adaptive for the environment of the organism.

During this lab exercise you will determine the optimal values of temperature, pH, substrate concentration, and enzyme concentration for a common enzymatic reaction, the hydrolysis of starch by **amylase.**

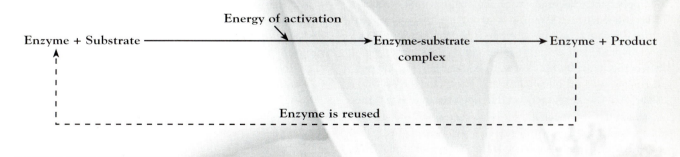

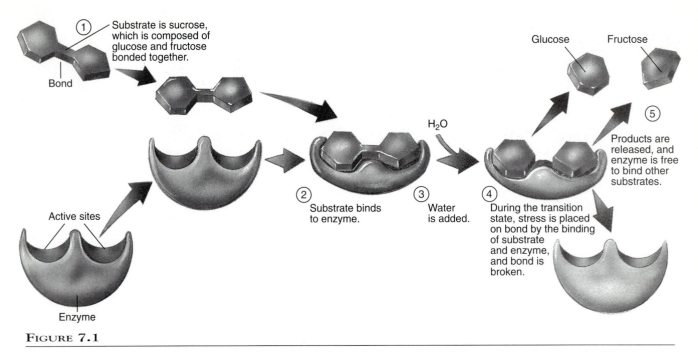

FIGURE 7.1

The catalytic cycle of an enzyme. Enzymes increase the speed at which chemical reactions occur but are not themselves altered in the process. Here, the enzyme splits the sugar sucrose (steps 1, 2, 3, and 4) into its two parts, the simpler sugars glucose and fructose. After the enzyme releases the glucose and fructose, it is then ready to bind another molecule of sucrose (5) and begin the catalytic cycle again.

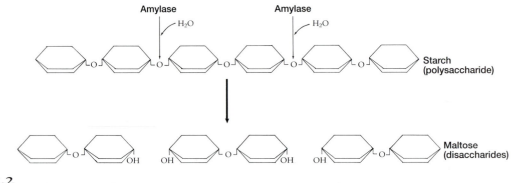

FIGURE 7.2

Hydrolysis of starch by amylase into subunits of maltose. Maltose is a disaccharide consisting of two molecules of glucose.

HYDROLYSIS OF STARCH BY AMYLASE

Starch is the primary energy-storage molecule of plants. It is a **polysaccharide** made of a long chain of **monosaccharides** called **glucose.** Amylase is an enzyme in plants and in other organisms, and its substrate is starch. Amylase is especially abundant in roots, tubers, and germinating seeds when the plant breaks down stored starch. The released sugars are used for metabolism and energy for the growing plant. Amylase is also produced by bacteria, and that is the source of the enzyme used in this lab exercise.

Amylase binds with a starch polymer and clips pairs of glucose subunits from its ends (fig. 7.2). The products of this enzymatic reaction are **disaccharides** called **maltose.** When amylase breaks the bond between adjacent glucose units a molecule of water is also split, and the OH^- and H^+ ions bind to the exposed ends of the broken starch polymer. This breaking of a chemical bond with the insertion of the ions of a water molecule is called **hydrolysis** (fig. 7.3), and in this manner amylase hydrolyzes starch. Hydrolysis is common, but some enzymes use other mechanisms to break chemical bonds.

In the laboratory you can easily follow the hydrolysis of starch by using iodine, which turns dark blue in the presence of starch (see exercise 2, "Biologically Important Molecules"). For this lab exercise, hydrolysis of starch by amylase is complete when a mixture of iodine and a sample of the solution does not turn blue. A positive (dark

7-2

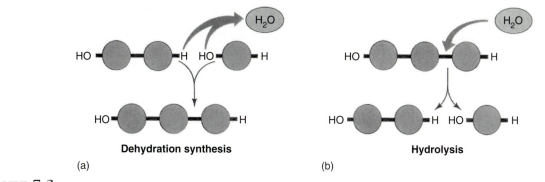

Dehydration synthesis

(a)

Hydrolysis

(b)

FIGURE 7.3

Dehydration synthesis and hydrolysis. (*a*) Biological molecules are formed by linking subunits. The covalent bond between subunits is formed in dehydration synthesis, a process that eliminates a water molecule. (*b*) Breaking such a bond requires the addition of a water molecule, a reaction called hydrolysis.

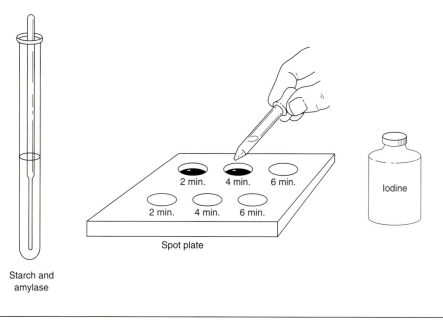

FIGURE 7.4

Setup of equipment used to detect the presence of starch.

blue) iodine-starch reaction indicates that starch is still present and the enzyme reaction is incomplete. Use procedure 7.1 to detect starch to monitor hydrolysis in each of your experiments.

Procedure 7.1
Detect starch and hydrolysis

1. Label a series of five to ten circles on a spot plate for an appropriate series of time intervals (e.g., 0 min, 2 min, 4 min, etc.). Place the spot plate on a sheet of white paper so you can judge the color easily.

2. Place a drop of iodine in each of the first few depressions of the spot plate.

3. Obtain a test tube containing the starch-amylase mixture to be tested.

4. Use a Pasteur pipet to extract 1 to 2 drops of the starch-amylase mixture from the test tube.

5. Release the drops onto the spot plate in one of the labeled circles containing iodine (fig. 7.4).

6. If the sample turns dark blue, then starch is present and the hydrolysis reaction is incomplete. Test again by repeating steps 4 to 5 at 2-minute intervals.

7. If the tested sample remains the color of iodine, then the reaction is complete and all the starch has been hydrolyzed. Note the time elapsed since the beginning of the reaction.

8. Record your results and wash your equipment.

EFFECT OF TEMPERATURE ON THE RATE OF HYDROLYSIS

Heat increases the rate of most chemical reactions. During enzymatic reactions faster molecular motion caused by heat increases the probability that molecules of enzyme will contact substrate. However, higher temperatures may alter the conformation of an enzyme and therefore do not always accelerate enzymatic reactions. Enzymatic reactions have an optimal range of temperatures; temperatures above or below this range decrease the reaction rate.

Question 1

a. What temperature range would you predict is

optimal for amylase? _____

Why? _____

b. Would you expect that the optimal temperatures for wheat, oyster, pig, or bacterial amylase are different?

_____

Why or why not? _____

c. Would you predict a narrower temperature range for

amylase activity in wheat or humans? _____

Why? _____

Procedure 7.2
Determine the effect of temperature on hydrolysis

1. Your instructor has prepared stock suspensions of starch and amylase. Pour 20 mL of the starch suspension into each of four test tubes; label the test tubes for temperatures of 5°, 20°, 40°, and 70°C.

2. Immerse the test tubes into four water baths adjusted to the temperature treatments. Leave the test tubes in the water baths for 10 minutes.

3. Using procedure 7.1 to detect starch and hydrolysis, test a few drops of solution from each test tube. This should verify the presence of starch.

4. Add 1 mL of amylase suspension to each test tube. Begin with the test tube at 5°C, and note the time when you add the enzyme to all four test tubes. Immediately after adding the enzyme, mix the contents of each test tube by stoppering and inverting the test tube. Keep the test tube in the water bath.

5. After adding the enzyme, use procedure 7.1 to test the contents of each test tube. Record your result on the data sheet provided at the end of this exercise. Repeat the procedure for each test tube at 2-minute intervals until the starch is completely hydrolyzed.

TABLE 7.1

HYDROLYSIS TIMES FOR THE ENZYMATIC BREAKDOWN OF STARCH BY AMYLASE AT DIFFERENT TEMPERATURES

Temperature	Hydrolysis Time
5 °C	
20 °C	
40 °C	
70 °C	

6. Record in table 7.1 the time for complete hydrolysis of starch for each of the four temperature treatments.

7. Wash your equipment when you have finished your work.

On the graph paper at the end of this exercise prepare a graph with the horizontal axis labeled *Temperature (°C)* and the vertical axis labeled *Time to Complete Hydrolysis (min)*. Be sure to title your graph. Plot the appropriate values from table 7.1 and connect the points.

Question 2

a. What temperature does your experiment indicate is

optimal for amylase? _____
Was your earlier prediction (in question 1) correct?

_____

b. Cold affects enzymes differently than does heat. How

do the effects differ? _____

_____

EFFECT OF pH ON THE RATE OF HYDROLYSIS

Enzymatic activity is sensitive to pH. Acidic and basic solutions are rich in H^+ and OH^- ions, respectively, and both of these ions are highly reactive (see plate II, fig. 3). Therefore, solutions with an extreme pH can change the conformation of a proteinaceous enzyme and alter its active site. Extreme pH can denature an enzyme just as drastically as do high temperatures.

Question 3

What pH do you predict is optimal for amylase? ____

_____

Why? _____

_____

TABLE 7.2

HYDROLYSIS TIMES FOR THE ENZYMATIC BREAKDOWN OF STARCH BY AMYLASE UNDER DIFFERENT CONDITIONS OF pH

pH	Hydrolysis Time
5	
6	
7	
8	

TABLE 7.3

HYDROLYSIS TIMES FOR ENZYMATIC BREAKDOWN OF STARCH BY AMYLASE UNDER DIFFERENT SUBSTRATE CONCENTRATIONS

Substrate Concentration	Hydrolysis Time
50%	
25%	
10%	
5%	

Procedure 7.3
Observe the effect of pH on hydrolysis

1. Your instructor has prepared four buffered solutions of pH 5, 6, 7, and 8. Label each of four test tubes with a different one of the pH values; then add 10 ml of the appropriate buffer to each.

2. Pipet 0.5 mL of amylase into each test tube. Stopper and invert the test tubes to mix the contents.

3. Using procedure 7.1 to detect starch and hydrolysis, test a few drops of solution from each test tube. This will verify the technique. You should not detect any starch.

4. Beginning with the test tube containing the buffer of lowest pH, add 10 mL of starch to each test tube. Stopper and invert the test tubes to mix the contents. Note the time.

5. After adding the starch, use procedure 7.1 to detect starch and hydrolysis to test for the presence of starch. Record your result on the data sheet provided at the end of the exercise. Repeat this procedure at 2-minute intervals for each test tube until hydrolysis is complete.

6. Record in table 7.2 the time for complete hydrolysis of starch for each of the four pH treatments.

7. Wash your equipment when you have finished your work.

On the graph paper at the end of this exercise prepare a graph with the horizontal axis labeled *pH* and the vertical axis labeled *Time to Complete Hydrolysis (min)*. Be sure to title your graph. Plot the appropriate hydrolysis times for each pH from table 7.2 and connect the points.

Question 4
What pH does your experiment indicate is optimal for amylase? _____
Was your earlier prediction (in question 3) correct?

_____

EFFECT OF SUBSTRATE CONCENTRATION ON THE RATE OF HYDROLYSIS

A high concentration of substrate increases the time needed for an enzyme to completely hydrolyze the substrate. It takes longer for a fixed amount of enzyme to hydrolyze more substrate. Although a high concentration of substrate may slightly enhance the reaction rate of some enzymes, high concentrations of substrate lengthen the time to hydrolyze all of the substrate.

Procedure 7.4
Determine the effect of substrate concentration on hydrolysis

1. Label four test tubes for the following substrate dilutions: 50%, 25%, 10%, and 5%.

2. In the four test tubes prepare the following starch dilutions:

Dilution	Starch	Water
50%	10 mL	10 mL
25%	5 mL	15 mL
10%	2 mL	18 mL
5%	1 mL	19 mL

3. Using procedure 7.1, test a few drops of solution from each test tube. This will verify the presence of starch.

4. Add 0.1 mL of amylase to each test tube and mix again. Note the time!

5. After adding enzyme to each test tube, use procedure 7.1 to test for the presence of starch. Record your result on the data sheet provided at the end of the exercise. Repeat this procedure at 2-minute intervals for each test tube to determine the time to complete hydrolysis for each substrate concentration.

6. Record in table 7.3 the time for complete hydrolysis for each of the four treatments.

7. Wash your equipment when you have finished your work.

On the graph paper at the end of this exercise prepare a graph with the horizontal axis labeled *Substrate Concentration (%)* and the vertical axis labeled *Time to Complete Hydrolysis (min)*. Be sure to title your graph. Plot the appropriate values from table 7.3 and connect the points.

Question 5

Was hydrolysis at a high concentration of substrate completed faster than hydrolysis at a low

concentration? _____

Why? _____

EFFECT OF ENZYME CONCENTRATION ON THE RATE OF HYDROLYSIS

If concentration of enzyme increases relative to the amount of substrate, then the rate of hydrolysis increases; that is, more enzyme hydrolyzes a substrate more quickly. Although the concentrations of enzymes in cells are usually low, the reactions are fast. A molecule of enzyme can catalyze the same reaction many times each second because it does not become part of the products. Enzymes are efficient under optimal conditions and are reused after each reaction. Notably, a cell can control the rate of its reactions by altering the concentrations of its enzymes. For example, if the substrate concentration increases dramatically, the cell can respond by making more enzyme.

Procedure 7.5
Determine the effect of enzyme concentration on hydrolysis

1. Label four test tubes for the following enzyme dilutions: 5.0%, 2.5%, 1.0%, and 0.5%.

2. In the four test tubes prepare the following enzyme dilutions:

Dilution	Amylase	Water
5.0%	2.0 mL	0.0 mL
2.5%	1.0 mL	1.0 mL
1.0%	0.4 mL	1.6 mL
0.5%	0.2 mL	1.8 mL

3. Add 18 mL of starch to each test tube. Note the time!

4. After adding starch to each test tube use procedure 7.1 to test for the presence of starch. Record your result on the data sheet provided at the end of the exercise. Repeat this procedure at 2-minute intervals for each test tube to determine the time required to completely hydrolyze the substrate at each enzyme concentration.

5. Record these times in table 7.4.

6. Wash your equipment when you have finished your work.

TABLE 7.4

HYDROLYSIS TIMES FOR ENZYMATIC BREAKDOWN OF STARCH BY AMYLASE UNDER DIFFERENT ENZYME CONCENTRATIONS

Enzyme Concentration	Hydrolysis Time
5.0%	
2.5%	
1.0%	
0.5%	

On the graph paper at the end of this exercise prepare a graph with the horizontal axis labeled *Enzyme Concentration (%)* and the vertical axis labeled *Time to Complete Hydrolysis (min)*. Be sure to title your graph. Plot the appropriate values from table 7.4 and connect the points.

Question 6

Did more enzyme result in faster completion of

hydrolysis in your experiment? _____

Why? _____

_____

Summary of Activities

1. Detect starch with iodine indicator.
2. Determine the effect of temperature on hydrolysis of starch and graph the results.
3. Determine the effect of pH on hydrolysis of starch and graph the results.
4. Determine the effect of substrate concentration on hydrolysis of starch and graph the results.
5. Determine the effect of enzyme concentration on hydrolysis of starch and graph the results.

Questions for Further Thought and Study

1. More substrate increases the probability that an enzyme will contact substrate and should increase the enzymatic reaction rate. How do you explain the increase in hydrolysis time when more substrate is present?

2. During your experiment, hydrolysis time should have decreased as enzyme concentration increased. Would a plateau in the hydrolysis time ever be reached as enzyme concentration continued to increase? Why or why not?

3. What term describes destruction of a protein's structure? What factors in addition to temperature influence a protein's structure?

Doing Botany Yourself

Design an experiment that uses a spectrophotometer to detect the progress and completion of hydrolysis of starch by amylase.

Writing to Learn Botany

Propose a mechanism involving enzyme production that would allow a cell to counteract a sudden increase in the amount of substrate.

Data Sheet

X = positive starch test
— = negative starch test

Temperature (°C)				pH				Subst. concen. (%)				Enzyme concen. (%)			
70°	40°	20°	5°	5	6	7	8	50	25	10	5	5.0	2.5	1.0	0.5
O	O	O	O	O	O	O	O	O	O	O	O	O	O	O	O
O	O	O	O	O	O	O	O	O	O	O	O	O	O	O	O
O	O	O	O	O	O	O	O	O	O	O	O	O	O	O	O
O	O	O	O	O	O	O	O	O	O	O	O	O	O	O	O
O	O	O	O	O	O	O	O	O	O	O	O	O	O	O	O
O	O	O	O	O	O	O	O	O	O	O	O	O	O	O	O
O	O	O	O	O	O	O	O	O	O	O	O	O	O	O	O
O	O	O	O	O	O	O	O	O	O	O	O	O	O	O	O
O	O	O	O	O	O	O	O	O	O	O	O	O	O	O	O
O	O	O	O	O	O	O	O	O	O	O	O	O	O	O	O
O	O	O	O	O	O	O	O	O	O	O	O	O	O	O	O
O	O	O	O	O	O	O	O	O	O	O	O	O	O	O	O
O	O	O	O	O	O	O	O	O	O	O	O	O	O	O	O
O	O	O	O	O	O	O	O	O	O	O	O	O	O	O	O
O	O	O	O	O	O	O	O	O	O	O	O	O	O	O	O
O	O	O	O	O	O	O	O	O	O	O	O	O	O	O	O
O	O	O	O	O	O	O	O	O	O	O	O	O	O	O	O
O	O	O	O	O	O	O	O	O	O	O	O	O	O	O	O
O	O	O	O	O	O	O	O	O	O	O	O	O	O	O	O
O	O	O	O	O	O	O	O	O	O	O	O	O	O	O	O
O	O	O	O	O	O	O	O	O	O	O	O	O	O	O	O
O	O	O	O	O	O	O	O	O	O	O	O	O	O	O	O
O	O	O	O	O	O	O	O	O	O	O	O	O	O	O	O
O	O	O	O	O	O	O	O	O	O	O	O	O	O	O	O
O	O	O	O	O	O	O	O	O	O	O	O	O	O	O	O

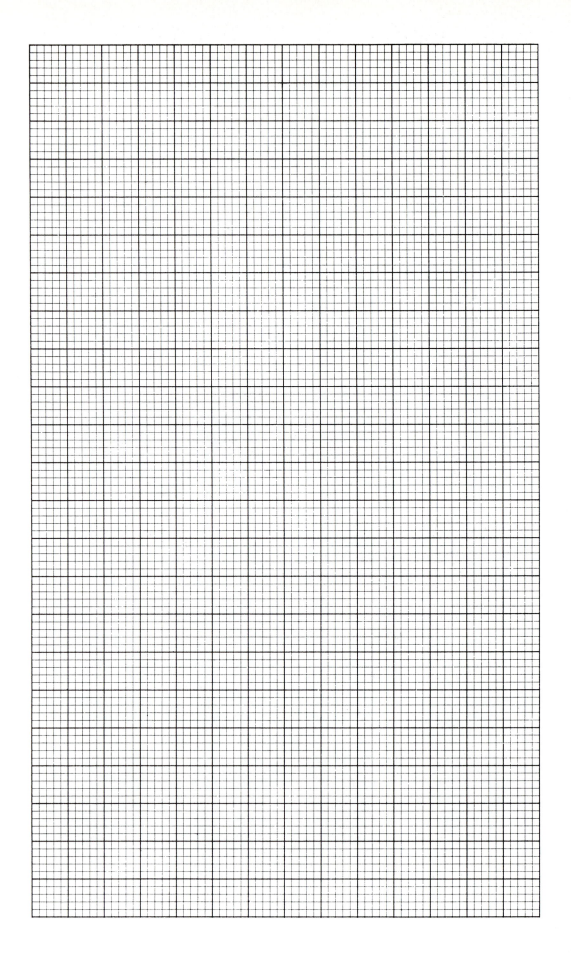

Respiration

Objectives

By the end of this exercise you will be able to

1. describe and interpret the results of experiments that measure respiration of yeast and plant cells

2. explain the difference between aerobic and anaerobic respiration.

Cellular respiration involves oxidation of organic molecules and a concomitant release of energy. Some of this energy is stored in chemical bonds of **adenosine triphosphate (ATP),** which is used later as a direct source of energy for cellular metabolism. Organisms use the energy stored in ATP to do work such as transport, synthesis, and reproduction.

In most cells, respiration begins with the oxidation of glucose to pyruvate via a set of chemical reactions called **glycolysis.** During glycolysis, energy released from each glucose molecule is stored in ATP. Glycolysis occurs with or without oxygen. If oxygen is present, most organisms continue respiration by oxidizing pyruvate to CO_2 and H_2O via chemical reactions of the **citric acid cycle** (tricarboxylic acid cycle, Krebs cycle). Organisms that use oxygen for respiration beyond glycolysis are called aerobes.

As aerobes oxidize pyruvate in the citric acid cycle, they store energy by reducing electron carriers such as NAD (nicotinamide adenine dinucleotide) and FAD (flavin adenine dinucleotide). These compounds later transfer their high-energy electrons to a series of compounds collectively called the electron transport chain. The **electron transport chain** generates proton gradients from energy stored in reduced NAD and FAD to form approximately eighteen-fold more ATP than that formed in glycolysis. Oxygen is the final electron acceptor in the electron transport chain. Without oxygen to accept electrons, the electron transport chain is not functional. The equations listed below compare the processes of aerobic and anaerobic respiration.

Other organisms called anaerobes live without oxygen and may even be killed by oxygen in their environment. These organisms perform glycolysis, but the pyruvate resulting from glycolysis is reduced via fermentation to either CO_2 and ethanol (in plants and some microbes) or lactic acid (in other microbes and oxygen-stressed muscles of animals). We can summarize anaerobic respiration (fermentation) in the following equations:

Summary Equation for Respiration

$$C_6H_{12}O_6 \ + \ 6\,O_2 \ \longrightarrow \ 6\,CO_2 \ + \ 6\,H_2O \ + \ \text{Energy as ATP} \ + \ \text{Heat}$$

glucose   oxygen   carbon dioxide   water

Anaerobic Respiration in Plants and Some Microbes

$$C_6H_{12}O_6 \ \longrightarrow \ 2\,C_2H_5OH \ + \ 2\,CO_2 \ + \ \text{Energy as ATP} \ + \ \text{Heat}$$

glucose   ethanol   carbon dioxide

Anaerobic Respiration in Animals and Other Microbes

$$C_6H_{12}O_6 \ \longrightarrow \ 2\,CH_3CHOHCOOH \ + \ \text{Energy as ATP} \ + \ \text{Heat}$$

glucose   lactic acid

Anaerobic respiration does not involve the citric acid cycle or electron transport chain, which both produce additional ATP. Thus, the ability of an organism to live in the absence of oxygen comes at a price: Anaerobic respiration produces many fewer ATP per glucose molecule than does aerobic respiration.

Question 1

a. Why must organisms take in oxygen and release CO_2?

_____

b. What is the advantage of anaerobic respiration? ____

_____

In today's exercise, you will study some of the major features of cellular respiration. Let's begin with a type of anaerobic respiration with which some of you are already familiar: alcoholic fermentation by yeast.

PRODUCTION OF CO_2 DURING ANAEROBIC RESPIRATION

Yeast are fungi used in baking and producing alcoholic beverages. They can respire in the absence of O_2 and can oxidize glucose to ethanol and CO_2. To demonstrate CO_2 production during anaerobic respiration by yeast (fermentation), follow procedure 8.1. In this experiment you will observe the effects of four compounds on respiration.

Pyruvate—the product of glycolysis; pyruvate is reduced to ethanol or lactic acid during fermentation

$MgSO_4$—provides Mg, which is a cofactor that activates some enzymes of glycolysis

NaF—an inhibitor of some enzymes of glycolysis

Glucose—a common organic molecule used as an energy source for respiration

Procedure 8.1

Demonstrate CO_2 production during anaerobic respiration

1. Label six fermentation tubes and add the solutions listed in table 8.1.

2. Completely fill the remaining volume in fermentation tubes 1 to 5 with the yeast suspension provided.

3. Fill the remaining volume in tube 6 with distilled water. Your instructor will demonstrate how to tip the fermentation tube to remove air bubbles.

4. Cover the outlet of each tube with a piece of plastic film or insert a foam plug.

5. Make a pinhole in the film to release pressure.

6. Incubate the tubes at 40°C; check for the production of CO_2 bubbles every 10 minutes for 40 minutes.

7. After 40 minutes, measure the height (in millimeters) of the bubbles of accumulated CO_2. Record your results in table 8.1.

Question 2

a. In your experiment which of the following compounds—sodium pyruvate, magnesium sulfate, and sodium fluoride—inhibited respiration? _____

_____

Why? _____

b. Which promoted respiration? _____

Why? _____

c. How did the concentration of NaF affect respiration in your experiment? _____

_____

d. Which compounds in the previous list are an intermediate in the respiratory pathway? _____

_____

e. Why did tube 5 produce CO_2 even though an inhibitor of glycolysis was present? _____

_____

f. Did magnesium (a cofactor that activates many enzymes) promote respiration? _____

If not, what are some possible reasons? _____

_____

g. What was the purpose of tube 6? _____

_____

h. Smell the contents of the tube containing the most CO_2. What compound do you smell? _____

_____

_____

TABLE 8.1

EFFECTS OF FOUR COMPOUNDS ON CO_2 PRODUCTION DURING ANAEROBIC RESPIRATION

Tube	Compound					Height of CO_2 Bubbles after 40 Minutes (mm)
	3 M Na Pyruvate	0.1 M MgSO$_4$	0.1 M NaF	5% Glucose	Water	
1	—	—	—	2.5 mL	5.0 mL	_____
2	—	5.0 mL	—	2.5 mL	—	_____
3	—	—	0.5 mL	2.5 mL	4.5 mL	_____
4	—	—	5.0 mL	2.5 mL	—	_____
5	2.5 mL	—	2.5 mL	2.5 mL	—	_____
6	2.5 mL	—	—	2.5 mL	2.5 mL	_____

i. What is the economic importance of fermentation by yeast? _____

_____

j. What gas is responsible for the holes in baked bread?

_____

_____

OXYGEN CONSUMPTION DURING AEROBIC RESPIRATION

Aerobic respiration uses oxygen as the terminal electron acceptor in the electron transport chain and reduces the oxygen to water. You can measure aerobic respiration by measuring the consumption of oxygen. During respiration, CO_2 is produced while O_2 is consumed. In the following experiment, KOH is used to absorb the CO_2. Therefore, the net change in gas volume is a measure of oxygen consumption.

Procedure 8.2

Determine oxygen consumption during aerobic respiration

1. Fill one test tube half full with germinating pea seeds and another half full with heat-killed pea seeds.
2. Cover the contents of each tube with a loose-fitting plug of cotton.
3. Cover the cotton with approximately 1 cm of loosely packed pellets of potassium hydroxide (KOH) (fig. 8.1).
4. Place a stopper containing a capillary tube with an attached outlet tube into both tubes containing peas (fig. 8.2).

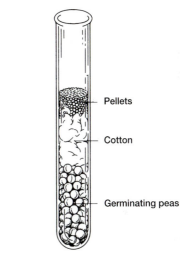

FIGURE 8.1

Test tube containing seeds, cotton, and KOH pellets.

5. Vertically clamp the tubes to a ring stand so that the bottom of each tube is submerged in a room-temperature water bath. The water bath will minimize temperature fluctuations in the tube.
6. Use a Pasteur pipet to inject enough dye into each capillary tube so that approximately 1 cm of dye is drawn into each capillary tube.
7. After waiting 1 minute for equilibration, attach a pinch clamp to the outlet tube and mark the position of the dye with a wax pencil.
8. Use a wax pencil to mark the position of the dye every 5 minutes for the next 15 minutes.
9. After each time interval measure the distance the fluid moved from its starting point; record your data in table 8.2.

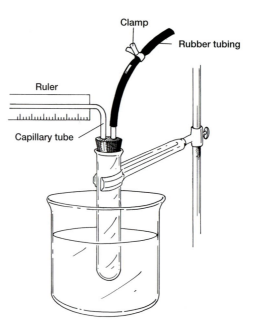

Clamp

Rubber tubing

Ruler

Capillary tube

FIGURE 8.2

Test tube containing stopper with capillary tubes attached.

TABLE 8.2

OXYGEN CONSUMPTION BY SEEDS AT THREE TEMPERATURES

Treatment	0 min	Distance Moved by Fluid (mm)		
		5 min	10 min	15 min
Room temperature	0			
Ice bath	0			
Warm water bath	0			

10. Remove the pinch clamp from the outlet valve and return the dye to the end of the capillary tube by tilting the capillary tube.

11. Repeat steps 1 to 10 using tubes incubated in an ice bath and a warm water bath. Record your results in table 8.2.

Question 3

a. In which direction did the dye move? _____

b. Why? _____

_____

c. What was the purpose of adding heat-killed seeds to

one tube? _____

d. What does this experiment tell you about the influence of temperature on oxygen consumption

during cellular respiration? _____

_____

PRODUCTION OF CO₂ DURING AEROBIC RESPIRATION

CO_2 produced during cellular respiration can combine with water to form carbonic acid:

$$CO_2 + H_2O \leftrightarrow H_2CO_3$$

In this experiment you will use phenolphthalein to detect changes in pH resulting from the production of CO_2 (and, therefore, carbonic acid) during cellular respiration. Phenolphthalein is red in basic solutions and colorless in acidic solutions. Thus, you can monitor cellular respiration by measuring acid production as change in pH. pH is a measure of the acidic or basic properties of a solution. A pH of 7 is neutral. Solutions having a pH < 7 are acidic, and solutions having a pH > 7 are basic.

In procedure 8.3 you will not directly measure the volume of CO_2 produced by a respiring organism. Instead, you will measure the volume of NaOH used to neutralize the CO_2 produced and thus calculate a relative measure of respiration.

KIMCHEE

Pickling is an ancient way of preserving food. Pickling involves the anaerobic fermentation of sugars to lactic acid. This acid lowers the pH of the medium, thereby creating an environment in which other food-spoiling organisms cannot grow. Common foods preserved with pickling include sauerkraut, yogurt, and dill pickles. The ancient Chinese cabbage product kimchee, still a major part of the Korean diet, is also made with pickling. Here's how to make kimchee.

1. Cut a 2-liter bottle just below the shoulder, as shown in figure 8.3*a*.

2. Add alternating layers of cabbage, garlic, pepper, and a sprinkling of salt in the bottle, pressing each layer down with a petri plate until the bottle is full (fig. 8.3*b*). If you've included chilis or pepper in your kimchee, do not touch your eyes or mouth.

3. Place the petri plate, rim side up, atop the ingredients. Press down. Within a few minutes, the salt will draw liquid from the cabbage, and that liquid will begin to accumulate in the bottle.

4. For the next hour or two, press the cabbage every 15 minutes or so. You should then be able to fit the bottle top inside the bottle to form a sliding seal (fig. 8.3*c*). When you press with the sliding seal, cabbage juice will rise above the petri plate and air will bubble out around the edge.

5. The cabbage will pack to half to two-thirds of the bottle's volume. Every day, press on the sliding seal to keep the cabbage covered by a layer of juice. What happens when you press on the cabbage? How do you explain this?

6. Use pH-indicator paper to measure and record the pH of the juice each day (fig. 8.3*d*).

7. After 4 to 7 days (depending on the temperature), the pH will have dropped from about 6.5 to about 3.5. Enjoy your kimchee!

(a)   (b)   (c)   (d)

FIGURE 8.3

Experimental setup for making kimchee. See the text for the recipe and procedure.

Procedure 8.3
Measure CO$_2$ production by *Elodea*

1. Obtain 225 mL of culture solution provided by your instructor. This solution has been dechlorinated and adjusted to be slightly acidic.

2. Place 75 mL of this solution in each of two beakers; one labeled "Control" and the other "*Elodea*."

3. Obtain some *Elodea* from your instructor. Determine the volume of the *Elodea* by the following procedure:

   a. Place exactly 25 mL of water in a 50-mL graduated cylinder.

TABLE 8.3

CO₂ PRODUCTION BY ELODEA DURING RESPIRATION

Treatment	Volume of Elodea (mL)	Milliliters of NaOH (mL NaOH)	Relative Respiration rate for Elodea (mL NaOH)	Respiration Rate per mL of Elodea (mL NaOH/mL organism)
Elodea beaker				
Control beaker	0		0	0

b. Place the *Elodea* in the cylinder and note the increase in volume above the original 25 mL. This increase equals the volume of the organism.

c. Record the volume in table 8.3. Gently place the *Elodea* in the beaker labeled *Elodea*.

4. Cover each beaker with plastic film. Place the *Elodea* beaker in the dark by covering it with a coffee can.

5. Allow the *Elodea* to respire for 15 minutes.

6. Remove the *Elodea* from the beaker and return it to its original culture bowl.

7. Add 1 drop of phenolphthalein to the contents of each beaker. The solutions should remain colorless because the solutions are acidic.

8. Obtain a buret to dispense NaOH (2.5 mL). Add NaOH drop by drop to the control beaker, thoroughly mixing the contents of the beaker after each drop. Record in table 8.3 the milliliters of NaOH required to reach the endpoint of phenolphthalein (i.e., make the solution pink).

9. Repeat step 8 for the beaker that contained the *Elodea*. Be sure to add NaOH only until the solution is the same shade of pink as the control. Record the number of milliliters of NaOH added to the *Elodea* beaker in table 8.3.

10. Determine the respiration rate for *Elodea* by subtracting the milliliters of NaOH added to the control beaker from the milliliters of NaOH added to the *Elodea* beaker. Record this value in table 8.3.

11. Determine the respiration rate per milliliter of *Elodea* by dividing the relative respiration rate by the volume of the *Elodea*. Record this value in table 8.3.

Question 4

a. Did respiration by *Elodea* reduce the pH of the surrounding solution? _____

b. How might your results change if you did not put the coffee can over the *Elodea* beaker? _____
_____

Summary of Activities

1. Determine the relative rates of anaerobic respiration by yeast subjected to different inhibitors and promoters.

2. Measure oxygen consumption by seeds.

3. Monitor CO₂ production by *Elodea*.

Questions for Further Thought and Study

1. Does cellular respiration occur simultaneously with photosynthesis in plants? How could you test your answer?

2. What role does cellular respiration play in a plant's metabolism?

3. What modifications of cellular respiration might you expect to find in dormant seeds?

Doing Botany Yourself

Repeat the steps to measure carbon dioxide production during anaerobic respiration. Incubate the tubes at 4°C (refrigerator), 20°C (incubator), and/or 50°C (incubator). How does temperature affect fermentation by yeast?

Doing Botany Yourself

Repeat procedure 8.3 to measure carbon dioxide production in *Elodea* using a beaker containing an animal such as a fish.

Photosynthesis

9

Objectives

By the end of this exercise you will be able to

1. relate each part of the summary equation for photosynthesis to the synthesis of a sugar
2. understand the light and biochemical reactions involved in photosynthesis
3. separate the photosynthetic pigments and measure photosynthesis.

Photosynthesis is unquestionably the most important series of chemical reactions that occurs on earth. Indeed, virtually all life depends on photosynthesis for food and oxygen. **Photosynthesis** is a complex chemical process that converts radiant energy (light) to chemical energy (sugars). A summary equation for photosynthesis as follows:

$$6\,CO_2 + 12\,H_2O \xrightarrow[chlorophyll]{light} (CH_2O)_n + 6\,H_2O + 6\,O_2$$

carbon water   sugars   water   oxygen
dioxide

Thus, photosynthesis is the light-dependent and chlorophyll-dependent conversion of carbon dioxide and water to sugars, water, and oxygen. Oxygen is released to

the environment, and the sugars are used either to fuel growth or are stored as **starch,** a polysaccharide. Notice that water is present on both sides of the summary equation; however, these are not the same water molecules. The "reactant" water molecules are split to release electrons during the photochemical reactions. The "product" water molecules are assembled from hydrogen and oxygen released during the photochemical and biochemical reactions.

Photosynthesis can be divided into two sets of reactions (fig. 9.1). Some characteristics of these reactions are compared here:

Photochemical Reactions	**Biochemical Reactions**
Light-dependent	Light-independent
Fast (practically instantaneous)	Slower, but still extremely fast
Splits water to release oxygen, electrons, and protons	Converts (fixes) carbon dioxide to glucose

In today's exercise, you'll investigate some of the major aspects of photosynthesis, beginning with isolation and identification of photosynthetic pigments.

PAPER CHROMATOGRAPHY OF PHOTOSYNTHETIC PIGMENTS

Light must be absorbed before its energy can be used. A substance that absorbs light is a **pigment.** The primary photosynthetic pigments that absorb light for photosynthesis are **chlorophylls *a*** and ***b*** (fig 9.2). Chlorophylls, however, are not the only photosynthetic pigments; **accessory pigments** such as **carotenes** and **xanthophylls** also absorb light and transfer it to chlorophyll *a*. The

light energy absorbed by accessory pigments is transmitted to chlorophyll (see plate VI, fig. 10).

Paper chromatography is a technique for separating dissolved compounds such as chlorophyll, carotene, and xanthophyll. When a solution of these pigments is applied to strips of paper, the pigments adsorb onto the fibers of the paper. When the tip of the paper is immersed in a solvent, the solvent is absorbed and moves up through the paper. As the solvent moves through the spot of applied pigments, the pigments dissolve in the moving solvent.

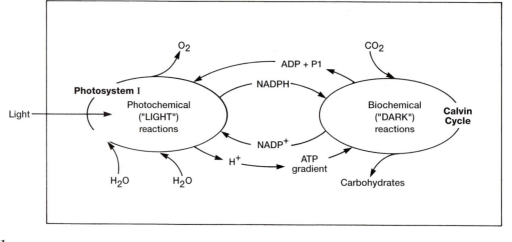

FIGURE 9.1

The photochemical and biochemical reactions of photosynthesis.

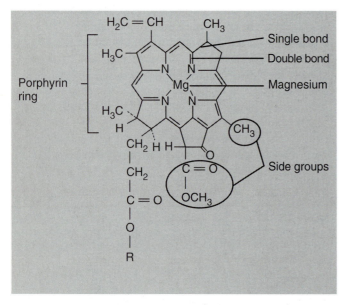

FIGURE 9.2

The chemical structure of chlorophyll *a*. The complex ring structure that forms the center of the molecule is called a porphyrin ring. Magnesium, a metal ion, lies at the center of the porphyrin ring. For the sake of simplicity, this diagram does not show the twenty-carbon alcohol that is attached to the chlorophyll *a* molecule; however, the position of this alcohol is marked by R.

However, the pigments can't always keep up with the moving solvent. Some pigments move almost as fast as the solvent, whereas others move more slowly. This differential movement of pigments results from each pigment's solubility and characteristic tendency to stick (i.e., be adsorbed) to the cellulose fibers of the paper. A pigment's molecular size, polarity, and solubility determine the strength of this tendency: Pigments adsorbed strongly move slowly, whereas those adsorbed weakly move fastest. Thus, each pigment has a characteristic rate of movement, and the pigments can be separated from each other.

In procedure 9.1, four bands of color will appear on the strip: a yellow band of carotenes, a yellow band of xanthophylls, a blue-green band of chlorophyll *a,* and a yellow-green band of chlorophyll *b.*

The relationship of the distance moved by a pigment to the distance moved by the solvent front is specific for a given set of conditions. We call this relationship the R_f and define it as follows:

$$R_f = \frac{\text{Distance moved by pigment}}{\text{Distance from pigment origin to solvent front}}$$

Thus, paper chromatography can be used to identify each pigment by its characteristic R_f. This R_f is constant for a given pigment in a particular solvent-matrix system.

Procedure 9.1

Separate plant pigments with paper chromatography

1. Observe the contents of the flask labeled "Plant Extract." You'll use paper chromatography to separate its pigments.

Question 1

What color is the plant extract, and why is it this

color? _____

_____

2. Obtain a strip of chromatography paper from your lab instructor. Handle the paper by its edges so that oil on your fingers does not contaminate the paper.

3. Make a faint pencil line across the paper approximately 2 cm from the tip of the paper. Now use a fine-tipped brush to apply a stripe of plant extract over the pencil mark (fig. 9.3). Blow the stripe dry and repeat this application at least ten times.

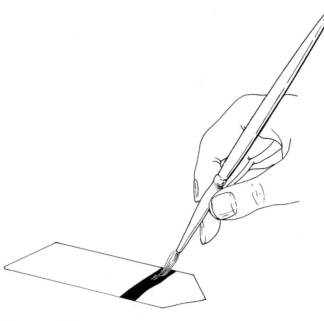

FIGURE 9.3

Application of pigment extract to chromatography strip.

FIGURE 9.4

Chromatography setup.

4. Place the chromatography strip in a test tube containing 2 mL of chromatography solvent (nine parts petroleum ether to one part acetone). Position the chromatography strip so that the tip of the strip (but not the stripe of plant extract) is submerged in the solvent. You can do this by hooking the strip of paper with a pin inserted in the tube's stopper (fig. 9.4).

> ### C A U T I O N
>
> The chromatography solvent is highly flammable, so extinguish all flames, hot plates, and cigarettes before you work with the solvent.

5. Place the tube in a test-tube rack and watch as the solvent moves up the paper.

6. Remove your chromatography strip before the solvent front reaches the top of the strip (after 5–10 minutes). Mark the position of the solvent front with a pencil and set the strip aside to dry. Observe the bands of color and draw your results on figure 9.5.

7. Measure the distance from the pigment origin to the solvent front and from the pigment origin to each pigment band. Calculate the R_f for each pigment; record your data in table 9.1.

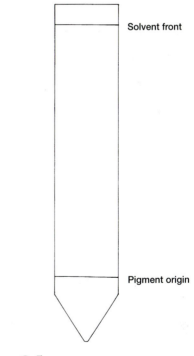

Solvent front

Pigment origin

FIGURE 9.5

Completed chromatogram.

TABLE 9.1

R_f VALUES FOR FOUR PLANT PIGMENTS

Pigment	R_f
Carotene	
Xanthophyll	
Chlorophyll a	
Chlorophyll b	

Question 2

a. What does a small R_f value tell you about the characteristics of the moving molecules? _____

_____

b. Which are more soluble in the chromatography solvent, xanthophylls or chlorophyll a? _____

_____

c. Would you expect the R_f value of a pigment to change if the composition of the solvent was altered?

_____

Why or why not? _____

_____

d. If yellow xanthophylls were present in the extract, why did the extract appear green? _____

_____

e. Is it possible to have an R_f value greater than 1? ____

_____

Why or why not? _____

_____

ABSORPTION OF LIGHT BY CHLOROPHYLL

A **spectroscope** is an instrument that separates white light into its component colors. These colors range from red to violet and appear as a spectrum when separated. Observe this spectrum by looking through the spectroscope provided in the lab. Now insert a chlorophyll sample between the light and spectroscope, and observe the resulting spectrum. Light not visible through the extract has been absorbed.

Question 3

As you view the chlorophyll sample through the spectroscope, what colors are diminished or absent?

_____

Based on this observation, complete the following absorption spectrum for chlorophyll. For each color, estimate the relative absorbance of that color by placing an X above the color name at the appropriate position along the y-axis. Connect the Xs for all colors to complete the absorption spectrum.

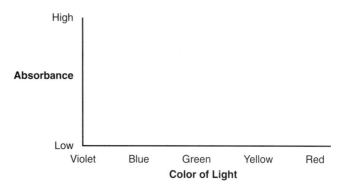

Question 4

What color of light is least effective for plant growth?

_____

Why? _____

FLUORESCENCE

Light reflected by a chlorophyll extract is green; hence the green color of the extract. However, not all light that strikes the extract is reflected; some is absorbed and reemitted at a longer wavelength. The wavelength of light that is re-emitted is determined by the structure of the molecule that is re-emitting the light. This phenomenon is **fluorescence.**

Place a tube of chlorophyll extract in front of a bright light. View the extract from the side.

Question 5

a. What color of light does the extract fluoresce? ____

_____

b. Could a chlorophyll extract fluoresce blue light? ____

_____

Why or why not? _____

ELECTRON TRANSPORT IN CHLOROPLASTS

The photochemical reactions of photosynthesis transfer electrons among various compounds. In 1937, Robin Hill demonstrated that isolated chloroplasts could transfer electrons in the absence of CO_2 if provided with an alternate or artificial electron acceptor. This observation indicated

TABLE 9.2

SOLUTIONS FOR COMPARISON OF PHOTOSYNTHETIC REACTION RATES

Tube	Chloroplasts	0.1 M PO$_4$ Buffer (pH 6.5)	H$_2$O	0.2 mM DCPIP
1	0.5 mL	3 mL	1.5 mL	0
2	0.5 mL	3 mL	0.5 mL	1 mL
3	0	3 mL	1.0 mL	1 mL
4	0.5 mL	3 mL	0.5 mL	1 mL

that electron transport does not require CO$_2$ fixation; that is, electron transfer and CO$_2$ fixation involve separate sets of reactions.

You can detect electron transfer using a dye called 2,6-dichlorophenol-indolephenol (DCPIP). In its oxidized state, DCPIP is blue. After accepting electrons, DCPIP becomes reduced and colorless. DCPIP can accept electrons released in chloroplasts during photosynthesis. The rate of DCPIP decoloration depends on its concentration and the rate of electron flow. By measuring the decoloration of DCPIP we can indirectly measure the rate of some reactions of photosynthesis. Because the rates of many chemical reactions are pH-dependent, a constant pH of approximately 6.5 is necessary for this experiment. The phosphate buffer used in this experiment maintains a constant pH of the incubation mixture.

Procedure 9.2
Observe electron transport in chloroplasts

1. Prepare test tubes according to table 9.2. Metabolically active chloroplasts will be provided by your instructor.

2. Mix the contents of each tube well and place tubes 1 to 3 approximately 0.5 m in front of a 100-watt bulb. Wrap tube 4 in aluminum foil, and place it with the other three tubes.

3. Observe the contents of the tubes intermittently, and notice the changes in color that you see.

4. If you have time, prepare a replicate of tube 2 in which water is replaced by 1 mL of 0.1 mM simizane or monuron, both of which are herbicides.

Question 6
a. What is the purpose of each of the four tubes used in this experiment? _____

_____

b. What happens when you illuminate the tube containing herbicide? _____

_____

c. Based on this result, what do you think is the mode of action of these herbicides? _____

_____

UPTAKE OF CARBON DIOXIDE DURING PHOTOSYNTHESIS

Phenol red (phenol-sulfonphthalein) is a pH indicator that turns yellow in an acidic solution (pH < 7) and becomes red in a neutral to basic solution (pH ≥ 7) (see plate II, fig. 3). In this experiment you will use phenol red to detect the uptake of CO$_2$ by a photosynthesizing plant. Recall that plants use CO$_2$ during the biochemical reactions of photosynthesis.

To detect CO$_2$ uptake you will put a plant into an environment that you have made slightly acidic with your breath. Carbon dioxide in your breath will dissolve in water to form carbonic acid, which lowers the pH of the solution.

$$H_2O \ + \ CO_2 \ \leftrightarrow \ H_2CO_3 \ \leftrightarrow \ H^+ \ + \ HCO_3^-$$

water   carbon   carbonic   hydrogen   bicarbonate
        dioxide   acid   ion   ion

As the plant fixes CO$_2$, the pH rises. When the pH rises above 7, the solution becomes red.

Procedure 9.3
Observe uptake of CO$_2$ during photosynthesis

1. Fill two test tubes half full with a dilute solution of phenol red provided by your laboratory instructor.

2. Blow slowly into each solution with a straw. Because excess carbonic acid will lengthen this experiment, stop blowing in the tubes as soon as the color changes to yellow.

3. Add pieces of healthy *Elodea* totaling about 10 cm to one tube. Pour off excess solution above the *Elodea*.

4. Place both tubes approximately 0.5 m in front of a 100-watt bulb for 30 to 60 minutes.

5. Observe the tubes about every 10 minutes.

Question 7

a. What happens to the color of the indicator? _____

_____

b. Considering the summary equation for photosynthesis, what is the basis for this change in color? _____

_____

USE OF LIGHT AND CHLOROPHYLL TO PRODUCE STARCH DURING PHOTOSYNTHESIS

Light-dependent reactions of photosynthesis occur on photosynthetic membranes. In photosynthetic bacteria, these membranes are the cell membrane itself (see fig. 4.2). In plants and algae, photosynthetic membranes are called **thylakoids,** which are located within a special organelle called a **chloroplast** (see fig. 4.3, plate V fig. 9c). Thylakoids are stacked to form columns called **grana.** Grana are held in place by **lamellae.** A semiliquid **stroma** bathes the interior of the chloroplast and contains the enzymes that catalyze the light-independent reactions of photosynthesis.

Sugars produced by photosynthesis are often stored as starch. Thus, starch production is another indirect measure of photosynthesis. To produce this starch, photosynthesis requires light as an energy source. In the absence of light, starch is not produced. Photosynthesis also requires chlorophyll to capture light energy. In the absence of chlorophyll, starch is not produced. In the following procedures, you will detect the presence of starch by staining it with a solution of iodine and demonstrate the requirement of light and chlorophyll for photosynthesis.

Procedure 9.4
Stain starch with iodine

1. Place separate drops of water, glucose, and starch solutions on a glass slide.
2. Add a drop of iodine to each and describe the result.

Procedure 9.5
Observe starch production during photosynthesis

1. Remove a leaf from a *Geranium* plant that has been illuminated for several hours.
2. After immersing the leaf in boiling water for 1 minute, bleach the pigments from the leaf by boiling the leaf in methanol for 3 to 5 minutes.

C A U T I O N

Exercise extreme caution when heating methanol.

---

3. Place the leaf in a petri plate containing a small amount of water, and then add 5 to 8 drops of iodine.
4. Observe any color change in the leaf.
5. Record in figure 9.6*a* the color of the leaves after each successive treatment.

Question 8

a. Was starch stored in the leaf? _____

b. Would you expect leaves to be the primary area for starch storage? _____

_____

Why or why not? _____

_____

Procedure 9.6
Observe the requirement of light for photosynthesis

1. Obtain a *Geranium* leaf that has been half or completely covered with metal foil for 3 to 4 days.
2. Repeat the bleaching and staining steps (2–3) described in the previous procedure.
3. Describe and explain any color change in the leaf.
4. Record in figure 9.6*b* the color of the leaves after each successive treatment.

Procedure 9.7
Observe the requirement of chlorophyll for photosynthesis

1. Obtain a leaf of a variegated *Coleus* plant and make a sketch of its pigmentation patterns.
2. Extract the various pigments and stain for starch according to the previous procedures.
3. Record in figure 9.6*c* the color of the leaves after each successive treatment.

Question 9

How does the pattern of starch storage relate to the distribution of chlorophyll?

Summary of Activities

1. Separate plant pigments by paper chromatography and calculate R_f values.
2. Observe the spectrum of white light after selective absorption by chlorophyll.
3. Graph the absorption spectrum of chlorophyll.

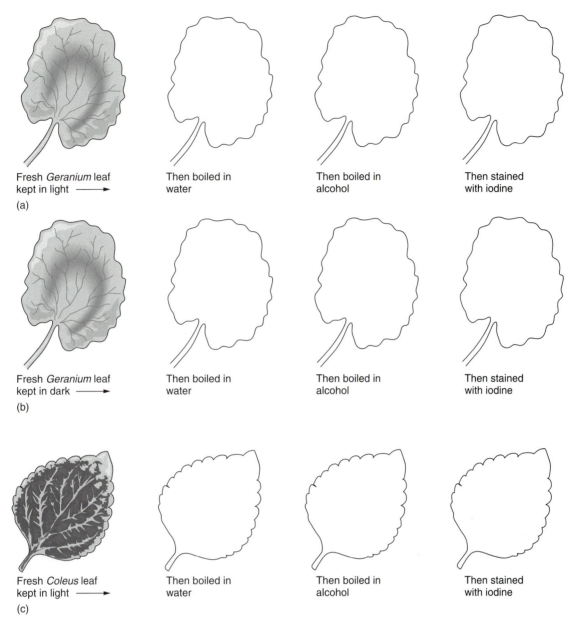

FIGURE 9.6

The requirement of light and chlorophyll and the production of starch during photosynthesis. Within each diagram, record the color of the leaf following the treatments to indicate (*a*) the production of starch, (*b*) the need for light, and (*c*) the need for chlorophyll for photosynthesis. Record your results from the appropriate procedure by writing the resulting color of each treated leaf directly onto the leaf outline.

4. Observe fluorescence.
5. Demonstrate the uptake of CO_2 during photosynthesis.
6. Stain starch with iodine.
7. Observe starch production during photosynthesis.
8. Demonstrate the requirement of light for photosynthesis.
9. Demonstrate the requirement of chlorophyll for photosynthesis.

Questions for Further Thought and Study

1. Why does chlorophyll appear green?
2. During darkness, what happens to starch formed in a leaf?
3. What causes leaves to turn from green to yellow and red in autumn?
4. Of what value to plants is starch?

5. What is the significance of electron transport in the photochemical (i.e., light dependent) reactions of photosynthesis?

6. During fluorescence, why is light reemitted at a longer wavelength?

C A U T I O N

Keep all solvents away from hot plates, flames, and other heat sources at all times. Extinguish all flames and hot plates before doing this experiment.

Doing Botany Yourself

Recall that respiration produces CO_2, which lowers the pH of a surrounding solution. Design an experiment to measure the relative dynamics (mass balance) of photosynthesis versus respiration for *Elodea*.

Writing to Learn Botany

Use a reference to determine the relative penetration of different wavelengths of light through water. Describe how this could affect the existence and distribution of submerged plants.

Genetics

Objectives

By the end of this exercise, you will be able to
1. understand simple genetic dominance, incomplete dominance, and lethal inheritance
2. explain the importance of Mendel's two laws
3. distinguish between an organism's phenotype and genotype
4. understand how transposons can affect the morphology and color of corn kernels.

Published papers are the primary means by which scientific discoveries are communicated (see appendix C). One of the most famous of these papers, titled "Experiments in Plant-Hybridization," was written in 1866 by Gregor Mendel, an Austrian monk. Although this paper later became the basis for genetics and inheritance, it went largely unnoticed until it was rediscovered independently by several European scientists in 1900. The experiments and conclusions in Mendel's paper now form the foundation of **Mendelian genetics,** the topic of this exercise.

Mendel's greatest contribution was to replace the blending theory of inheritance, which stated that all traits blended with each other, with the **particulate theory.** Mendel's particulate theory states that (1) inherited characters are determined by particular factors (now called genes), (2) these factors occur in pairs (genes on maternal and paternal homologous chromosomes), and (3) during formation of gametes via meiosis these genes segregate so that only one of the homologous pair is contained in a particular gamete. Each gamete has an equal chance of possessing either member of a pair of homologous chromosomes. This part of the particulate theory is collectively known as Mendel's first law, or the **law of segregation.** Mendel's second law, or the **law of independent assortment,** states that genes on nonhomologous or different chromosomes will be distributed randomly into gametes.

Before starting the exercise, briefly review some pertinent principles and terms. Be sure also to read the pages about genetics in your textbook.

A **gene** is a unit of heredity on a chromosome. A gene has alternate states called **alleles,** which are contributed to an organism by its parents. Alleles for a particular gene occur in pairs. Alleles that mask expression of other alleles but are themselves expressed are called **dominant;** these alleles are usually designated by a capital letter (e.g., T). Alleles whose expression is masked by dominant alleles are called **recessive,** and they are designated by a lowercase letter (e.g., t). The **genotype** of an organism includes all the alleles present in the cell, whether they are dominant or recessive. The physical appearance of the trait is the **phenotype.** Thus, if tallness (T) is dominant to dwarfness (t), a tall plant can have a genotype TT or Tt. A dwarf plant can only have the genotype tt. When the paired alleles are identical (TT or tt), the genotype is **homozygous. Heterozygous** refers to a pair of alleles that are different (Tt).

With this minimal review, let's now apply this information to solve some genetics problems.

SIMPLE DOMINANCE

Assume that tallness is dominant to dwarfness. If a homozygous tall plant is crossed (mated) with a homozygous dwarf plant, what will be the phenotype (physical appearance) and genotype of the offspring?

parents: TT (homozygous tall) $\times tt$ (homozygous dwarf)

gametes: T from one parent and t from the other parent

offspring: genotype = Tt

phenotype = tall

This first generation of offspring is called the **first filial,** or F_1, generation.

Each of the F_1 offspring (Tt) can produce two possible gametes, T and t, each with the same probability as heads or tails in a coin flip. Mendel noted that the gametes from each of the parents combine with each other randomly. Thus, we can simulate the random mating of gametes from the F_1 generation by flipping two coins simultaneously. Assume that heads designates the tall allele (T) and tails designates the dwarf allele (t). Flipping one coin will determine the type of gamete from one parent, and flipping the other will determine the gamete from the other parent. To simulate the results of 64 random matings, flip two coins simultaneously 64 times, and record the occurrence of each of the three possible combinations in table 10.1.

Question 1

What is the ratio of tall (TT or Tt) to dwarf (tt)

offspring? _____

Keeping these results in mind, now return to the original problem: What are the genotypes and phenotypes of the offspring of the F_1 generation?

parents: $Tt \times Tt$

gametes: (T or t) $\times$ (T or t)

offspring: TT, Tt, tT, tt (3 tall, 1 dwarf)

Thus, the theoretical genotypic ratio for the offspring of the F_1 generation is 1 TT:2 Tt:1 tt, and the phenotypic ratio is 3 tall:1 dwarf.

Question 2

a. How do these ratios compare with your data derived

from coin flipping?   _____

b. Would you have expected a closer similarity if you had flipped the coins 64,000 times instead of 64 times?

_____

Why or why not? _____

_____

Now solve the following problem.

Problem 1: The color of grains (karyopses) and height of corn plants are each often determined by a single gene. Examine the following materials provided by your instructor: (1) ears of corn with red and yellow grains, and (2) a tray of tall and dwarf plants. Record your observations, and determine the probable genotypes of the parents of each cross.

Color of Corn Grains

Number of red grains _____

Number of white grains _____

Ratio of red to white grains _____

Probable genotypes of parents _____ $\times$ _____

TABLE 10.1

RESULTS OF COIN-FLIPPING EXPERIMENT SIMULATING RANDOM MATING OF HETEROZYGOUS (Tt) INDIVIDUALS

Results of Flips	Number of Occurrences
Heads-heads = TT (tall)	
Heads-tails = Tt (tall)	
Tails-tails = tt (dwarf)	

Height of Plants

Number of tall plants _____

Number of dwarf plants _____

Ratio of tall to dwarf plants _____

Probable genotypes of parents _____ $\times$ _____

The crosses in this problem involve only one trait and are thus called **monohybrid crosses.** Let's now examine a cross involving two traits; that is, a **dihybrid cross.** Your lab instructor will review with you the basis for working genetics problems involving dihybrid crosses. In corn, red seed color (R) is dominant to white seed color (r), and smooth (S) is dominant to wrinkled seed(s). Observe the cobs of corn derived from a cross between parents with genotypes of $RRSS$ and $rrss$.

Question 3

What ratios of red to white and smooth to wrinkled offspring would you predict for the cross of $RRSS \times rrss$?

_____

To test your prediction in question 3, count the number of kernels for five rows for each of the following phenotypes:

red, smooth   _____

red, wrinkled   _____

white, smooth   _____

white, wrinkled   _____

Question 4

a. How do the test data compare with your prediction

in question 3? _____

b. What ratios of red to white and smooth to wrinkled offspring would you predict for the cross of

$RrSs \times RrSs$? _____

_____

Repeat these counts for cobs derived from self-crosses of the F_1 generation.

red, smooth _____

red, wrinkled _____

white, smooth _____

white, wrinkled _____

Question 5

How do these data compare with those that you

predicted in question 4b? _____

_____

INCOMPLETE DOMINANCE

Some traits such as flower color are controlled by incomplete dominance. In this type of inheritance, the heterozygous genotype results in an intermediate characteristic. For example, if a plant with red flowers (RR) is crossed with a plant with white flowers (rr), all of the offspring in the first filial (F_1) generation will have pink flowers (Rr).

parents: RR (red) $\times rr$ (white)

gametes: $R \times r$

offspring: Rr (pink)

Question 6

What are the expected ratios of red, pink, and white flowers in a cross involving two pink-flowered

parents? _____

LETHAL INHERITANCE

Lethal inheritance is the inheritance of a gene that kills the offspring. Observe the tray of green and albino seedlings of corn. The albino plants cannot photosynthesize and therefore die as soon as their food reserves are exhausted.

Question 7

a. What is the ratio of green to albino seedlings? _____

_____

b. Based on this ratio, what might you expect were the

genotypes of the parents? _____

c. Why is it impossible to cross a green plant and an

albino plant? _____

_____

TRANSPOSONS

For much of this century, geneticists thought that genes do not move in cells. However, in 1947 Barbara McClintock proposed that genes could move within and between chromosomes. McClintock based her conclusion on a series of experiments involving genetic crosses in corn. Specifically, McClintock showed that there is a fragment of DNA that can move to and be inserted at the locus for the production of pigments in corn kernels. Because this insertion renders the cell unable to make the purple pigment, the resulting kernel is yellow or white. However, subsequent removal of the DNA fragment results in the cell resuming production of the purple pigment; therefore, the resulting kernel is purple. Thus, Indian corn often has kernels with varying pigmentation, depending on when the DNA fragment was inserted or removed. A similar phenomenon occurs with the production of other pigments in corn kernels. The translocation to and from the locus for production of these pigments several times during kernel development produces the red-orange swirls characteristic of many kernels of Indian corn.

The fragments of DNA that McClintock studied are now called **transposons.** Transposons are a useful tool for genetic engineering because they provide a way of inserting foreign DNA into a host cell's chromosome. For her work McClintock received the Nobel Prize in 1983.

Procedure 10.1
Observe corn kernels to understand the effects of transposons

1. Work in groups of two to four. Obtain an ear of Indian corn for your group.
2. Look for examples of kernels with (a) purple or white spots, (b) red-orange swirls, and (c) other unusual color patterns.
3. Use the information presented in this exercise and in your textbook to determine how transposons could produce such unusual patterns of pigmentation.

Summary of Activities

1. Review the fundamental terms of genetics.
2. Simulate random mating (pairing) of two alleles by dual coin flips.
3. Solve problems involving simple dominance and monohybrid crosses.
4. Solve a problem involving incomplete dominance.
5. Determine the effects of transposons on the morphology and color of corn kernels.

Questions for Further Thought and Study

1. What determines how often a phenotype occurs in a population?

2. Is it possible to determine the genotype of an individual with a dominant phenotype? How?

3. Why is hybrid seed so expensive to produce?

4. Are dominant characteristics in a population always more frequent than recessive characteristics? Why or why not?

Writing to Learn Botany

Organisms that are heterozygous for a trait are often called carriers of that trait. What does this mean?

Mitosis

11

Objectives

By the end of this exercise you will be able to

1. describe events associated with the cell cycle
2. describe events associated with mitosis
3. distinguish the phases of mitosis on prepared slides of mitotic cells
4. stain and examine chromosomes in mitotic cells.

Cells grow, have specialized functions, and replicate during their lives. All of these activities are part of a repeating set of events called the **cell cycle.** A major feature of the cell cycle is cellular replication, and a major feature of cellular

replication is mitosis. **Mitosis** is replication of the nucleus in eukaryotic cells. Eukaryotic cells, found in plants, animals, fungi, and protists, have a membrane-bounded nucleus, so when a cell replicates the nucleus also replicates. In contrast, prokaryotic cells lack nuclei and do not undergo mitosis.

Mitosis is usually associated with cytokinesis, the division of the cell and cytoplasm into two halves, each containing a new nucleus. In certain tissues of plants, cytokinesis is delayed or does not occur at all, and the cells are multinucleate. Mitosis and cytokinesis are important because they provide a mechanism for orderly growth of living organisms.

THE CELL CYCLE

This exercise emphasizes events associated with mitosis, but mitosis is only part of the cell cycle (fig. 11.1). The remainder of the cycle is called the **interphase** and is subdivided further into **gap 1** (G_1), **synthesis** (S), and **gap 2** (G_2) phases.

The cell cycle begins with formation of a new cell and ends with replication of that cell. The G_1 phase occurs after mitosis and cytokinesis and includes a variety of growth processes and synthesis of compounds other than DNA. During the following S phase, DNA that makes up the chromosomes is duplicated. At the end of the S phase each chromosome consists of an identical pair of chromosomal DNA strands called **chromatids** that are attached at a **centromere** (fig. 11.2). A chromosome can be generally defined as a centromere and the DNA-protein complex attached to it, whether in one or two strands (two chromatids). During the G_2 phase, molecules and structures necessary for mitosis are synthesized. Other events probably remain to be discovered. Mitosis follows G_2 and usually lasts for less than 10% of the time of the cell cycle

but may vary greatly in length. Cell cycles of 10 to 30 hours are common. Cytokinesis may begin during mitosis but is highly variable in length and timing.

Not all plant cells continue to divide and complete the cell cycle. The most rapidly dividing or possibly the only dividing cells of a mature plant are in specialized regions called **meristems** at the growing tips of roots, shoots, and other specialized areas. Other cells in leaves and certain tissues of stems and roots halt their cell cycle and become specialized.

Question 1

a. Mitosis is often referred to as "cellular division." Why is it more accurately called nuclear replication? _____

_____

_____

b. Does the cell cycle have a beginning and an end? __

_____

Explain. _____

_____

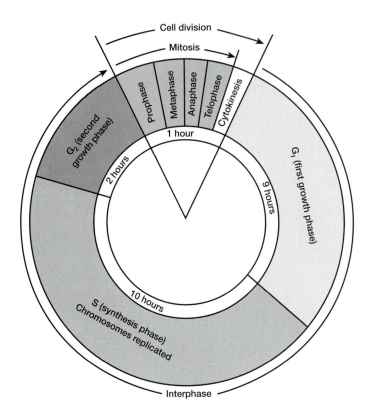

FIGURE 11.1

The cell cycle.

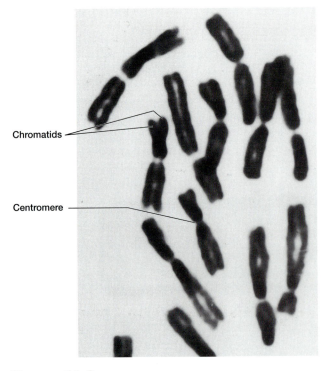

Chromatids

Centromere

FIGURE 11.2

Plant chromosomes.

MITOSIS

Mitosis (1) separates the genetic material that was duplicated during interphase into two identical sets of chromosomes, and (2) reconstitutes a nucleus to house each set. As a result, mitosis produces two identical nuclei from one.

Mitosis is traditionally divided into four stages: **prophase, metaphase, anaphase,** and **telophase** (fig. 11.3). The actual events of mitosis are not discreet but occur in a continuous sequence; separation of mitosis into four stages is merely convenient for our discussion and organization. During these stages, important cellular structures are synthesized and perform the mechanics of mitosis. For example, before prophase begins, microtubules and actin filaments form a narrow bundle, called a **preprophase band,** just inside the plasma membrane and around the nucleus. The band's plane is perpendicular to the movement of chromosomes during mitosis. The new cell wall will form along the plane of this band.

After the preprophase band appears the spindle apparatus forms. The **spindle apparatus** includes **spindle fibers** made of microtubules and is elliptical with its ends pointing towards opposite poles of the cell. Some of the fibers attach to each chromosome's **kinetochore,** a complex of proteins that binds to the centromere. As fibers attach,

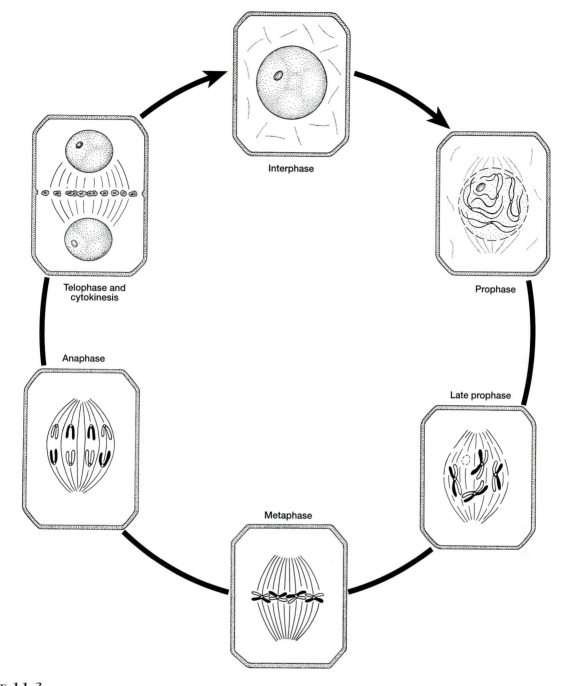

FIGURE 11.3

Stages of mitosis in plants.

chromosomes distribute themselves on the spindle apparatus and are subsequently moved and separated to opposite poles. The newly arrived chromosomes at each pole will eventually become part of two new nuclei. This mitotic distribution of chromosomes also occurs if the cell is haploid (has a single set of chromosomes); the vegetative cells of many organisms such as fungi are haploid rather than diploid (have a double set of chromosomes). However, the steps of mitosis are the same as for diploid cells.

CYTOKINESIS

After mitosis is completed (or near its end), cytokinesis begins. A system of microtubules, called a **phragmoplast,** forms on either side of the plane once formed by the preprophase band and the old metaphase plate. Dicytosome vesicles with cell wall precursors accumulate and fuse along this plane. The phragmoplast, fused vesicles, and

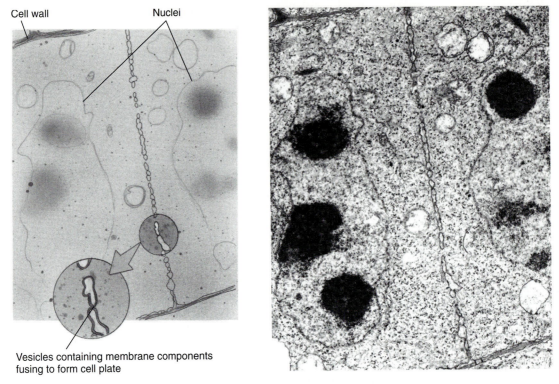

Cell wall   Nuclei

Vesicles containing membrane components
fusing to form cell plate

FIGURE 11.4

Cytokinesis in plant cells. In this photograph and companion drawing, a cell plate is forming between daughter nuclei. Once the plate is complete there will be two cells.

new wall material form the **cell plate** (fig. 11.4). The cell plate grows and new cell walls form within it to separate the two new cells.

Procedure 11.1
Describe the specific events of mitosis

Before your lab meeting, review in your textbook the events associated with each stage of mitosis and with the preparatory stage, or interphase. Describe these events in table 11.1. This list can serve as an excellent study guide, so be as complete as possible. One event for each stage is provided as an example. Some events and structures occur only in plant cells, whereas others occur only in animal cells. Mark these events in your list with an asterisk.

Question 2
a. If a nucleus has eight chromosomes during interphase, how many chromosomes will it have

during metaphase? _____

b. How many chromosomes will it have after mitosis is

complete? _____

Understanding the movements of chromosomes is crucial to understanding mitosis. You can simulate these movements easily with chromosome models made of pipe

cleaners or popsicle sticks. This is a simple exercise but a valuable one. It will be especially helpful when you are comparing the events of mitosis to the events of meiosis, which you will simulate in the next exercise.

Procedure 11.2
Simulate chromosomal replication and movement during mitosis

1. Examine the materials to be used as chromosome models provided by your instructor.

2. Identify the differences in chromosomes represented by various colors, lengths, or shapes of materials. Also identify materials representing centromeres.

3. Place a sheet of notebook paper on your lab table to use in representing the boundaries of the mitotic cell.

4. Assemble the chromosomes needed to represent nuclear material in a cell of a diploid organism with a total of six chromosomes. Place the chromosomes in the cell.

5. Arrange the chromosomes to depict the position and status of chromosomes during interphase G_1. (During G_1 the chromosomes are usually not condensed, as the chromosome models imply, but the models are an adequate representation.)

TABLE 11.1

Interphase (Includes G_1, S, G_2)	Each chromosome is replicated.
Prophase	Mitotic apparatus is formed.
Metaphase	Chromosomes align on the metaphase plate or equatorial plane of the spindle apparatus.
Anaphase	Chromatids are physically separated.
Telophase	Mitotic apparatus is disassembled.

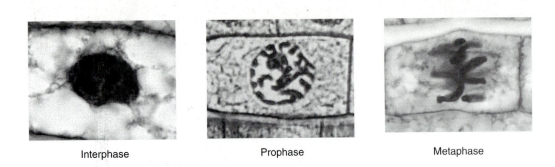

Interphase — Prophase — Metaphase

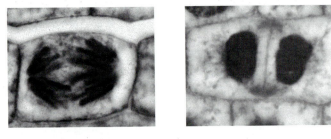

Anaphase — Telophase

FIGURE 11.5

Mitosis in onion root tip.

6. Depict the status of chromosomes after completing interphase S. Use additional "nuclear material" if needed.

7. Depict the chromosomes during prophase.

8. Depict the chromosomes during metaphase.

9. Depict the chromosomes during anaphase.

10. Depict the chromosomes during telophase.

11. Draw the results of cytokinesis and the re-formation of nuclear membranes.

12. Chromosomal events occur as a continuous process of movements rather than in distinct steps. Therefore, repeat steps 4 to 11 as a continuous process and ask your instructor to verify your simulation.

Our plant model to study cellular replication in plants is the root tip of *Allium* (onion). The root tip has a meristem specialized for rapid cell division.

Procedure 11.3
Observe and describe mitosis in *Allium*

1. Examine a prepared slide of a longitudinal section through an onion (allium) root tip.

2. Search for examples of all stages of mitosis (fig. 11.5).

3. Search for signs of cell plate formation.

4. In figure 11.6, diagram a plant cell with a diploid number of three pairs of chromosomes in each of the stages of mitosis. Recall that **diploid** refers to a nucleus with two of each type of chromosome. Be

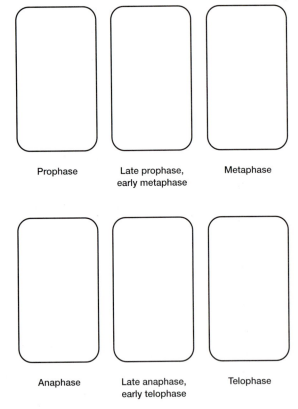

Prophase — Late prophase, early metaphase — Metaphase

Anaphase — Late anaphase, early telophase — Telophase

FIGURE 11.6

Stages of mitosis in plants.

TABLE 11.2

DURATION OF EACH STAGE OF MITOSIS IN AN ONION ROOT CELL

Stage of Mitosis	Predicted Duration (hours)	Number of Cells in Each Stage	Class Total for Each Stage	Calculated Duration (hours)
Interphase				
Prophase				
Metaphase				
Anaphase				
Telophase				
Totals				

sure to label the cell wall and cell plate. Two extra cells are provided for you to diagram midphases to demonstrate the continuity of the events of mitosis.

5. Prepared cross sections of cells show only two dimensions, but mitosis is a three-dimensional process. In the space below draw two cells in metaphase: one in which the cross section is parallel to the axis of the spindle apparatus and one in which the cross section is perpendicular to the spindle apparatus.

Question 3

a. What region of a root has the most mitotic activity?

b. Why is pinching of the cytoplasm inadequate for cytokinesis in plant cells? _____

c. Locate a plant cell in late telophase. What is the volume of the two new cells relative to a mature cell?

TIME LAPSED DURING STAGES OF CELL REPLICATION

The length of the cell cycle in actively dividing root tips of *Allium* is approximately 24 hours. With this in mind, predict the amount of time that a cell spends in each stage. Record your estimates in table 11.2.

Procedure 11.4
Calculate duration of each stage of mitosis

1. Reexamine a prepared slide of an onion root tip and count all the cells in one field of view. Count the number of cells in each of the stages of mitosis.

2. Repeat this procedure for other fields of view until you count more than 100 cells.

3. Record your results along with the combined results of your class in table 11.2.

4. Now calculate the time spent in each stage based on a 24-hour cell cycle by dividing the number of cells in each stage by the total number of cells counted. Then multiply this fraction by 24. Record your results in table 11.2.

Although your calculations are only a rough approximation of the time spent in each stage, they clearly illustrate the differences in duration of the various stages of mitosis.

Question 4

a. Why are the combined data from all the class members more meaningful than your results alone?

b. How close were your predictions for the length of each stage of mitosis? _____

c. What sources of error can you list for this technique to determine the time elapsed during the stages of mitosis? _____

PREPARING AND STAINING CHROMOSOMES

Your instructor has prepared some living onion root tips for you to process further and use to observe the stages of mitosis.

Procedure 11.5
Stain chromosomes

1. Obtain an onion root tip and place it in a small vial with Schiff's reagent for 30 minutes. Handle Schiff's reagent carefully because it is a colorless liquid that becomes bright red after reaction. Keep the vial in the dark and at room temperature until the tip becomes purple. Your instructor may have already stained some root tips for you.

2. Place the root tip in a drop of 45% acetic acid on a slide and cut away all except the terminal 1 mm of the tip.

3. Crush the root tip with a blunt probe and cover the tissue with a coverslip.

4. Smash the tissue by pressing on the coverslip with the eraser of your pencil. Your instructor will demonstrate this procedure.

5. Scan your preparation at low magnification to locate stained chromosomes. Then switch to high magnification and locate formations of chromosomes that indicate each of the stages of mitosis.

6. Add a drop of acetic acid to the edge of the coverslip to avoid desiccation.

7. Locate as many stages of mitosis as you can. Be sure to look at preparations done by other students.

Summary of Activities

1. List the major events of each stage of mitosis.
2. Simulate the replication and movement of chromosomes during mitosis.
3. Observe mitosis in onion cells.

4. Diagram cells in metaphase from three-dimensional perspectives.
5. Diagram a plant cell with three pairs of chromosomes in each stage of mitosis.
6. Calculate the duration of each stage of mitosis.
7. Stain chromosomes in an onion root tip.

Questions for Further Thought and Study

1. Interphase has sometimes been called a "resting stage." Why is this inaccurate?

2. Most metabolism not associated with mitosis occurs during G_1 of interphase. What events occur during other phases of mitosis that might inhibit general metabolism?

3. Review prokaryotic cellular replication in your textbook. List the fundamental differences between it and eukaryotic cellular replication.

4. Some specialized cells such as neurons and red blood cells lose their ability to replicate when they mature. Which phase of the cell cycle do you suspect is terminal for these cells? Why?

Writing to Learn Botany

Refer to your textbook to review the properties of a chemical called colchicine. Describe the effects that colchicine treatment would have on dividing cells. What is the mechanism of its effect?

Meiosis

By the end of this exercise you will be able to

1. describe the events of meiosis
2. describe similarities and differences between meiosis and mitosis
3. understand the most significant events of meiosis
4. discuss the relevance of meiosis to sexual reproduction.

Meiosis is a process of nuclear division that reduces by half the number of chromosomes in the resulting cells. Thus, meiosis is sometimes called "reduction division." For many organisms the resulting cells become gametes. In plants, the resulting cells are spores that grow into the haploid stage of a plant's life cycle.

In organisms that reproduce sexually, chromosomes typically are **diploid** (2N); that is, chromosomes occur as double sets (pairs) in each nucleus. The two chromosomes of a pair are called **homologous chromosomes,** and each homolog of a pair has the same sites, or **loci,** for the same genes. Meiosis reduces the number of chromosomes to a **haploid** (1N), or single, set. This reduction is important because a cell with a haploid set of chromosomes can fuse with another haploid cell during sexual reproduction and restore the original, diploid number of chromosomes to the new individual. In addition to reducing the number of chromosomes, meiosis shuffles the genetic material so that each resulting cell carries a new and unique set of genes.

Question 1

a. Why would shuffling genetic material to produce new combinations of characteristics be advantageous to a species? _____

_____

b. When would it be harmful? _____

_____

Meiosis, like mitosis, is preceded by replication of each chromosome to form two chromatids attached at a centromere. However, reduction of the chromosome number and production of new genetic combinations result from two events that do not occur in mitosis. First, meiosis includes two rounds of chromosome separation (known as meiosis I and meiosis II). Chromosomes are replicated in the first round but not in the second round. Thus, the genetic material is replicated once and divided twice. This produces half the original number of chromosomes.

Second, early in meiosis each chromosome (composed of two chromatids) pairs along its length with its homolog. This pairing of homologous chromosomes is called **synapsis,** and the four chromatids exchange various segments of genetic material (fig. 12.1). This exchange of genetic material, called **crossing-over,** produces new genetic combinations. During crossing-over, genetic material is not gained or lost, but afterward each chromatid of the chromosomes contains different genetic information.

Question 2

a. Synapsis occurs after chromosomal DNA has replicated. How many chromatids are involved in crossing-over of a homologous pair of

chromosomes? _____

_____

b. Suppose synapsis occurred between two homologous chromosomes, and one had alleles for white petals and narrow leaves, whereas the other had alleles for red petals and broad leaves. How many different combinations of these alleles would

be possible? _____

_____

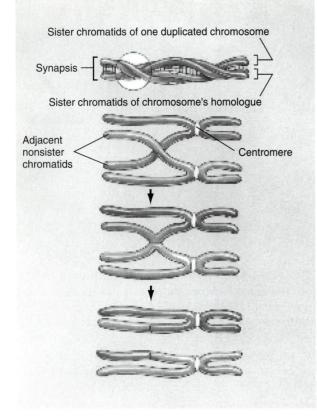

FIGURE 12.1

Crossing-over occurs when paired chromosomes exchange arms. The open circle highlighting the upper chromosome indicates a region where tightly bundled, homologous chromosomes are pairing. The lower sequence of chromosomes shows expanded views of nonsister chromatids exchanging segments of chromosomes.

STAGES AND EVENTS OF MEIOSIS

Although meiosis is a rather continuous process, we can study it more easily by dividing it into stages just as we did for mitosis. Indeed, meiosis and mitosis are similar, and their corresponding stages of prophase, metaphase, anaphase, and telophase have much in common. However, meiosis takes longer than mitosis because meiosis involves two nuclear divisions instead of one. These two divisions are called meiosis I and meiosis II. The chromosome number is reduced (through reduction division) during meiosis I, and chromatids composing each chromosome are separated in meiosis II. Each division involves the events of prophase, metaphase, anaphase, and telophase (fig. 12.2).

Before your lab meeting, review in your textbook the events associated with each stage of meiosis, including the preparatory stage, interphase. List these events in

table 12.1. This list can serve as an excellent study guide, so be as complete as possible. Ask your instructor to check for errors.

Premeiotic Interphase

Meiosis I is preceded by an interphase that is similar to the G_1, S, and G_2 of mitotic interphase, including replication of the chromosomes. Each chromosome is replicated.

Compare the outline in table 12.1 with your outline in the previous exercise on mitosis (table 11.1).

Question 3
a. If a nucleus has eight chromosomes when it begins meiosis, how many chromosomes does it have after

 telophase I? _____

 telophase II? _____

b. What are the major differences between the events

 of meiosis and mitosis? _____

 _____

c. What are some minor differences, and why do you

 consider them minor? _____

 _____

To understand meiosis you must understand the movements of chromosomes. You can easily simulate these movements with chromosome models made of pipe cleaners or Popsicle sticks. This is a simple but valuable exercise. It is especially instructive if you compare your simulation of meiosis with your simulation of mitosis from the previous lab exercise. The chromosomal events are significantly different.

Procedure 12.1
Simulate the replication and movement of chromosomes during meiosis

1. Examine the materials to be used as chromosome models provided by your instructor.

2. Identify the differences in chromosomes represented by various colors, lengths, or shapes of the materials. Also identify materials representing centromeres.

3. Place a sheet of notebook paper on your lab table. Use it to represent the boundaries of the meiotic cell.

4. Assemble the chromosomes needed to represent the nuclear material in a cell of a diploid organism with a total of six chromosomes (three homologous pairs). Place the chromosomes in the cell.

5. Arrange the chromosomes to depict the position and status of chromosomes during interphase G_1. (During

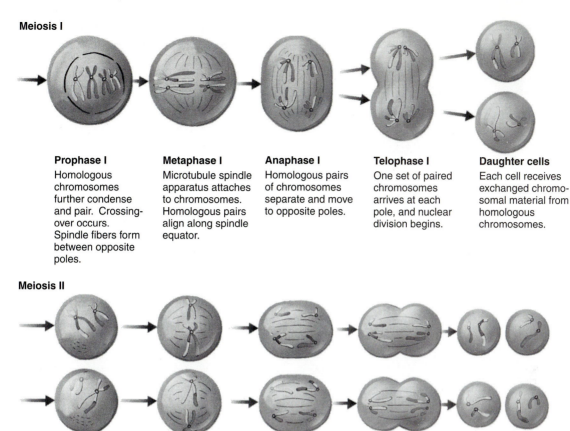

Prophase I
Homologous chromosomes further condense and pair. Crossing-over occurs. Spindle fibers form between opposite poles.

Metaphase I
Microtubule spindle apparatus attaches to chromosomes. Homologous pairs align along spindle equator.

Anaphase I
Homologous pairs of chromosomes separate and move to opposite poles.

Telophase I
One set of paired chromosomes arrives at each pole, and nuclear division begins.

Daughter cells
Each cell receives exchanged chromosomal material from homologous chromosomes.

Meiosis II

Prophase II
Chromosomes recondense. Spindle fibers form between opposite poles.

Metaphase II
Microtubule spindle apparatus attaches to chromosomes. Chromosomes align along spindle.

Anaphase II
Sister chromatids separate and move to opposite poles.

Telophase II
Chromatids arrive at each pole, and cell division begins.

Daughter cells
Cell division complete. Each cell ends up with half the original number of chromosomes.

FIGURE 12.2

Stages of meiosis. Meiosis involves two serial nuclear divisions with no DNA replication between them. Meiosis produces four daughter cells, each with half the original amount of DNA. Crossing-over occurs in prophase I of meiosis.

G_1 the chromosomes are usually not condensed as the chromosome models imply; nevertheless, the models are an adequate representation.)

6. Depict the chromosomes after completing interphase S. Use additional "nuclear material" if needed.

7. Depict the chromosomes during prophase I, during metaphase I, during anaphase I, and during telophase I.

8. Draw the results of cytokinesis, which occurs at this stage in some organisms.

9. Depict the status of chromosomes during prophase II for both daughter nuclei. Repeat this for metaphase II, anaphase II, and telophase II.

10. Draw the results of cytokinesis and the re-formation of nuclear membranes.

11. Chromosomal events are a continuous process rather than distinct steps. Therefore, repeat steps 4 to 10 as a continuous process and ask your instructor to verify your simulation.

In flowering plants, meiosis occurs in the anthers and ovary of the flower. In the anther, the spores resulting from meiosis produce a stage of the life cycle (pollen) that will eventually produce male gametes. In the ovary the resulting spores produce a stage of the life cycle (ovule) that will eventually produce female gametes. You'll learn more about these events in exercise 28. For this exercise you will observe prepared slides showing stages of the beginning, middle, and end of meiosis I and II in a representative plant.

TABLE 12.1

EVENTS OF MEIOSIS

Prophase I	Homologous chromosomes line up side by side.
Metaphase I	Paired homologous chromosomes are aligned.
Anaphase I	Microtubules of spindle fibers contract.
Telophase I	Each spindle pole has one member of each homologous pair.
Prophase II	Spindle apparatus is formed.
Metaphase II	Chromosomes align on spindle apparatus.
Anaphase II	Chromatids are separated.
Telophase II	Meiotic apparatus is disassembled.

12-4

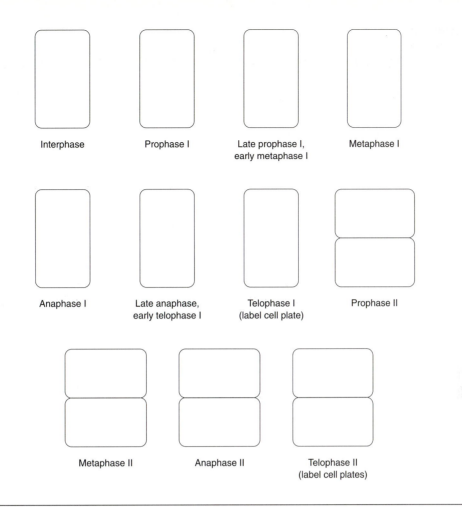

Interphase Prophase I Late prophase I, early metaphase I Metaphase I

Anaphase I Late anaphase, early telophase I Telophase I (label cell plate) Prophase II

Metaphase II Anaphase II Telophase II (label cell plates)

FIGURE 12.3

Stages of meiosis.

Procedure 12.2
Observe and diagram stages of meiosis

1. Examine the following prepared slides of stages of meiosis in a *Lilium* anther (see fig. 30.5).

    a. *Lilium* anther—early prophase I

    b. *Lilium* anther—late prophase I

    c. *Lilium* anther—first meiotic division

    d. *Lilium* anther—second meiotic division

    e. *Lilium* anther—pollen tetrads. Each of these cells will produce a pollen grain.

2. Examine the following prepared slides of stages of meiosis in a *Lilium* ovary (see fig. 30.8).

    a. *Lilium* ovary—"mother cell," prophase I

    b. *Lilium* ovary—binucleate stage, end of meiosis I

    c. *Lilium* ovary—four-nucleate stage, end of meiosis II

3. In figure 12.3, diagram a plant cell with three pairs of chromosomes in each of the stages of meiosis. Be sure to label the cell wall and cell plate. Two extra cells are provided for you to diagram midphases to demonstrate the continuity of the events of meiosis.

MITOSIS VERSUS MEIOSIS

Mitosis and meiosis are both forms of cellular replication, but they play different roles in the life cycle of plants. Mitosis may occur in either haploid or diploid cells and is necessary for cell production and growth. Meiosis occurs in diploid cells. Its role is to produce cells with a reduced number of chromosomes and shuffle the genetic material so an organism can reproduce sexually. Review table 12.2, which further contrasts mitosis and meiosis.

TABLE 12.2

CONTRASTING FEATURES OF MITOSIS AND MEIOSIS

Mitosis	Meiosis
Performed by all cells at some stage in their development	Performed only by specialized germinal cells
Produces two cells	Produces four cells
Includes one nuclear division	Includes two nuclear divisions
Resulting cells are diploid	Resulting cells are haploid
Resulting cells are genetically identical	Resulting cells are genetically different
No pairing of homologous chromosomes or crossing-over	Includes pairing of homologous chromosomes and crossing-over

Summary of Activities

1. Outline the events of each phase of meiosis.
2. Simulate chromosomal replication and movement during meiosis.
3. Examine the stages of meiosis in *Lilium* anthers and ovaries.
4. Contrast mitosis and meiosis.

Questions for Further Thought and Study

1. Would evolution occur without the events of meiosis and sexual reproduction? Why or why not?
2. Which process, meiosis or mitosis, is most accurately referred to as nuclear division? Why?

Writing to Learn Botany

Wouldn't it be easier for the chromosomes in a cell simply to divide once rather than duplicate and then divide twice during meiosis? Why do you suppose this does not occur?

13

DNA Isolation and Bacterial Transformation

Objectives

By the end of this exercise you will be able to

1. isolate DNA from a bacterium
2. understand how temperature and pH affect DNA structure
3. insert a gene for resistance to ampicillin into a bacterium

During the past decade we have witnessed a biological revolution whose impact could ultimately rival that of the Industrial Revolution. This revolution is based on molecular biology and biotechnology (genetic engineering), which allow us, for the first time, to directly manipulate an organism's DNA and genetic fate.

The biotechnological revolution began in 1973, when Stan Cohen and Herb Boyer produced the first chimeric organism. This organism wasn't the mythological combination of a lion, goat, and serpent; rather, it was a harmless bacterium. Specifically, Cohen and Boyer transplanted a gene for antibiotic resistance from a frog into a bacterium and thus "engineered" an antibiotic-resistant organism. Biologists quickly realized the significance of Cohen and Boyer's work and began trying to apply genetic engineering to humans and human concerns. A major breakthrough occurred in 1980, when molecular biologists succeeded for the first time in inserting a human gene for interferon, an antiviral drug, into a bacterium. When the "transformed" bacterium reproduced, generating billions of progeny, it became a miniature drug factory. As a result, biologists could cheaply mass-produce a drug that was previously expensive and generally unavailable. Similar bioengineering projects have mass-produced other drugs (e.g., insulin) and have profoundly affected our society. Indeed, molecular biology is now used in forensics (e.g., to link suspects to crimes, to settle paternity disputes), medicine (e.g., gene therapy, drug production), hiring practices (e.g., pinpointing employees at high risk for cancer), and the production of new foods and environmentally benign pesticides.

In this exercise you'll learn two techniques used routinely by molecular biologists: (1) isolation of DNA and (2) genetic transformation.

ISOLATION OF DNA

Isolation of DNA is a routine and important procedure for molecular biologists. Once isolated, the DNA and the sequence of its subunits can be determined, manipulated, or altered. Bacteria contain only about 10^{-14} g of DNA, which accounts for approximately 5% of the organism's dry weight. However, molecules of this DNA can be very long; for example, the DNA in an *E. coli,* if strung out, would be approximately 1 mm long. (By analogy, if the bacterium were the size of a grapefruit, then its DNA would be more than 80 km long.) The ability to pack this much DNA into a tiny cell is impressive, especially since DNA is a rather stiff molecule (fig. 13.1).

In this exercise, you will isolate DNA from *Halobacterium salinarum*, a halophilic ("salt-loving") bacterium that grows only in salty environments (4–5 M NaCl). It's especially easy to isolate DNA from this organism because its cell walls disintegrate when placed in low-salt environments (0–2 M NaCl).

Procedure 13.1
Isolate DNA from *Halobacterium*

1. Several days ago your laboratory instructor prepared petri plates of culture media and inoculated them with *Halobacterium salinarum*. Obtain one of these cultures from your instructor.

A human chromosome contains an enormous amount of DNA. The dark element at the bottom of the photograph is the protein matrix of a single chromosome. All of the surrounding material is the DNA of that chromosome.

2. Use a flexible plastic ruler to scrape the bacterial growth from the surface of the agar and collect it in a test tube.
3. Add 1 mL of distilled water to the tube.
4. Place a small piece of plastic film over the top of the test tube and hold it securely in place with your thumb. Invert the tube several times to mix the contents. This mixing with water lyses the cells, and the liquid will become viscous (thick and resistant to flow).
5. Use an eyedropper to slowly and gently pour 1 mL of ice-cold 95% ethanol down the side of the tube. Do not shake or disturb the tube. DNA will precipitate at the interface between the layers of water and ethanol.
6. Insert a glass rod into the liquid. Rotate the rod. DNA will adhere to the rod as it is twirled.

Question 1

What is the texture of the DNA you've isolated?

_____

The DNA that you've isolated is not pure; it is contaminated with small amounts of protein and RNA.

THE INFLUENCE OF HEAT AND pH ON DNA

Heat and pH strongly affect the properties of DNA. For example, DNA typically denatures at alkaline pH and at 80 to 97°C. Test these effects with the following procedures.

Procedure 13.2
Test the influence of heat on DNA

1. Precipitate DNA in a test tube following steps 1 to 5 in procedure 13.1 for isolating DNA.
2. Place the tube into a boiling water bath for 10 minutes.
3. Place the tube in an ice bath.
4. Insert and twirl a glass rod in the tube.
5. Compare the viscosity of the heat-treated DNA with untreated DNA.

Question 2
a. What effect does heat have on the viscosity of DNA?

_____

b. What do you think is the mechanism for this change

in viscosity? _____

Procedure 13.3
Test the influence of pH on DNA

1. Precipitate DNA in a test tube following steps 1 to 5 in procedure 13.1 for isolating DNA.
2. Add 2 mL of 1.0 N NaOH to the tube.
3. Insert and twirl a glass rod in the tube.
4. Compare the viscosity of the alkali-treated DNA with untreated DNA and with heat-treated DNA.

Question 3
a. What effect does alkaline pH have on the viscosity

of DNA? _____

b. What do you think is the mechanism for this change

in viscosity due to pH? _____

_____

b. Which affects DNA viscosity more, heat or alkalinity?

_____

GENETIC TRANSFORMATION

Much of biotechnology is based on **genetic transformation,** which is the uptake and expression of DNA by a living cell. A successful transformation, summarized in figure 13.2, requires three conditions: (1) a host into which DNA can be

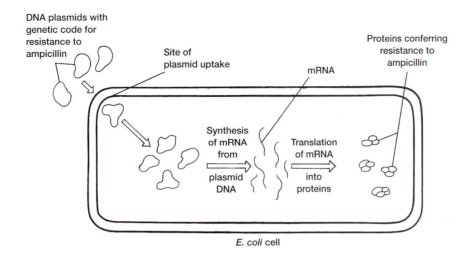

FIGURE 13.2

Transformation of an *E. coli* cell with plasmid DNA. In this example, the DNA plasmid contains the genetic code for resistance to the antibiotic ampicillin. After uptake of the plasmid, the code is transcribed to messenger RNA, which is translated during protein synthesis.

inserted, (2) a means of carrying the DNA into the host, and (3) a method for selecting and isolating the organisms that were successfully transformed.

The Host: *Escherichia coli*

The **host organism** you'll use is *Escherichia coli*, one of the most intensively studied organisms in the world. *E. coli* has two properties that make it ideally suited for transformation:

- It contains only one chromosome made of 5 million base pairs. This is less than 0.2% of that of the human genome.

- *E. coli* grows rapidly. Transformations in bacteria are rare and occur in only ~0.1% of cells. Therefore, transformations are observed most easily in large, rapidly growing populations. *E. coli* is ideal for transformation studies because, in ideal conditions, it divides every 20 minutes. As a result, in 10 hours a bacterium can produce a billion progeny (thirty generations) in only 1 mL of nutrient broth.

Only a small percentage of bacterial cells in a culture can be transformed. Also, small lengths of DNA are taken up more readily than long lengths. However, **competence** of the bacteria (i.e., the chances for successful transformation) increases during the early and middle stages of its growth. Competence also increases when cells suspended in a cold solution of $CaCl_2$ are heat-shocked. Yield is usually ~10^6 transformants per milligram of DNA available for insertion.

A Vector to Move DNA into the Host

A biological vector is a DNA molecule that carries DNA sequences into a host. The simplest bacterial vectors are plasmids, which are circular pieces of DNA made of 1,000

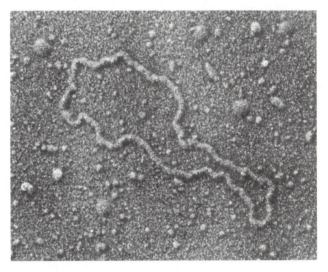

FIGURE 13.3

A famous plasmid. The circular molecule in this electron micrograph was the first plasmid used successfully to clone a vertebrate gene, *pSC101*. Its name refers to it being the 101st plasmid isolated by Stanley Cohen.

to 200,000 base pairs (fig. 13.3). Plasmids exist separately from the bacterial chromosome, and they must contain a gene that confers some selective advantage (e.g., resistance to an antibiotic) to remain in the host. We don't completely understand how plasmids enter host cells, but they seem to enter consistently.

Selecting Transformed Organisms

You'll insert into *E. coli* a plasmid (+P) containing a gene for resistance to ampicillin, an antibiotic lethal to many bacteria. Then you'll select transformed bacteria based on their resistance to ampicillin by spreading the transformed organisms onto nutrient medium containing ampicillin. Organisms that grow on this medium have been transformed. Because *E. coli* grows so fast, you can check for transformed organisms only 12 to 24 hours after completing the experiment.

Procedure 13.4
Transform E. coli

1. Carefully read this procedure and review fig. 13.4 before beginning, and wash your hands. Your instructor will demonstrate the steps of the procedure that require sterile technique.

2. Fill a 250-mL beaker with about 50 mL of ethanol; place a glass-rod bacterial spreader in the ethanol to soak.

3. Obtain a test tube with 1 mL of sterile, yellow nutrient broth and a tube with 1 mL of clear, colorless 50 mM $CaCl_2$. Label these tubes NB and $CaCl_2$, respectively.

4. Fill a 250-mL beaker half full with crushed ice. Place the tube of $CaCl_2$ in the beaker of ice.

5. Obtain two sterile, plastic transformation tubes and label one of them (+)P and the other (−)P. Place them in the beaker of ice.

6. Obtain a packaged, sterile, plastic pipet and locate the graduation indicating a volume of 0.25 mL.

7. Open the packaged pipet without touching and contaminating the pipet's open end. Use this pipet to add 0.25 mL of a sterile, ice-cold solution of 50 mM $CaCl_2$ to each of the two transformation tubes. Use sterile technique.

8. Several days ago your lab instructor streaked plates of nutrient agar with *E. coli*. Scrape a colony (3 mm diameter) from one of these plates with a sterile, plastic inoculating loop. Use sterile technique.

9. Place the loop full of bacteria into the transformation tube labeled (−)P. Rinse the bacteria from the loop by gently twirling the loop handle between your fingers.

10. To mix the bacterial suspension, open a sterile, packaged pipet. Insert the pipet into the suspension in the bottom of the tube and gently suck the fluid in and out of the tip three or four times. Be sure the tip is empty before withdrawing the pipet.

11. For tube (+)P, repeat steps 8 to 10. Use a fresh loop and pipet to inoculate and mix the bacteria. Try to get the same amount of bacteria into each tube. Replace the two tubes in the ice bath and chill the tubes for at least 5 minutes.

12. Use a sterile loop to obtain one loop full (10 μL) of an ice-cold solution of DNA plasmids from a vial kept by your instructor. Add this loop full of plasmids to tube (+)P and gently rotate the loop to rinse the plasmids from the loop. These plasmids contain the gene for resistance to ampicillin. Do *not* add plasmids to tube (−)P.

13. Place both tubes in ice for 15 minutes.

14. While the tubes are cooling, obtain two agar plates labeled −AM, meaning nutrient agar without ampicillin and two plates labeled +AM, meaning nutrient agar with ampicillin. Label one of each pair of plates as (−)P and the other two plates as (+)P.

15. Heat-shock the transformation tubes (−)P and (+)P by placing them in a 42°C water bath for 2 minutes. Heat shock increases the uptake of the plasmid by the bacterial cells.

16. Chill the tubes in ice for 5 minutes.

17. Remove the tubes from the ice bath; place them in a test-tube rack or empty beaker.

18. Using a sterile pipet, add 0.25 mL nutrient broth to each tube and tap gently to mix the contents.

19. Using sterile technique and a sterile pipet, transfer 0.10 mL of the (−)P cell suspension onto the agar's surface in the middle of the plate labeled +AM/(−)P. Transfer another 0.10 mL onto the plate labeled −AM/(−)P. Close the plates.

20. Using another sterile pipet, transfer 0.10 mL of the (+)P cell suspension onto the middle of the plate labeled +AM/(+)P. Transfer another 0.10 mL onto the plate labeled −AM/(+)P. Close the plates.

21. Light an alcohol lamp. Dip the bacteria spreader into the ethanol and then into the flame of an alcohol lamp. Let the ethanol burn away and count to ten to let the spreader cool. Before spreading the bacteria in the first of your four plates, further cool the spreader by touching it to the agar at the edge of the plate. Then touch the spreader to the cell suspension in the middle of the plate and gently drag it back and forth three times. Rotate the plate 90° and repeat. Remember to sterilize the bacteria spreader between each plate. Repeat this procedure to spread the bacteria on the other three plates.

22. Put your name and date on each of the four plates, and tape the plates together. Incubate the plates upside down at 37 °C.

23. Place all the tubes, loops, etc., in a central location for disposal. Wipe down your work area with a weak bleach solution and wash your hands before leaving the laboratory.

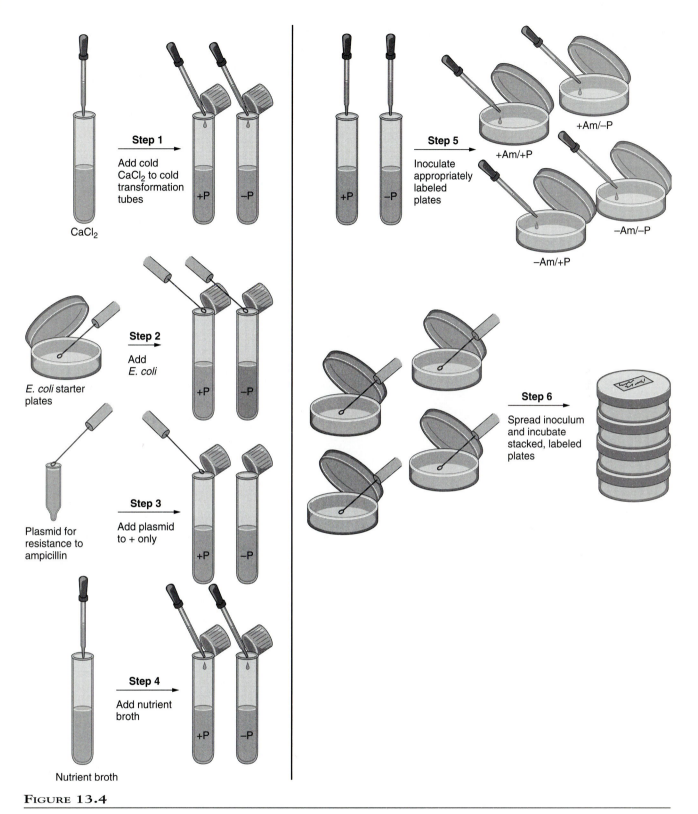

FIGURE 13.4

Summary of the procedure to transform bacteria by exposing *E. coli* to a plasmid.

24. On the circles below label the contents of the four plates. Indicate which plates had bacterial growth after 24 hours by drawing the appearance and coverage of bacterial colonies.

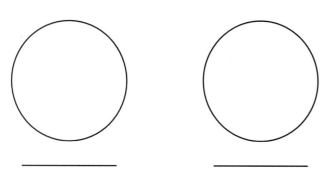

_____   _____

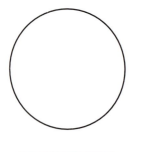 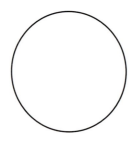

_____   _____

Question 4

a. Which treatment produced transformed bacteria?

   _____

b. How many transformed colonies grew on each plate?

   _____

c. What was the purpose of tube 2, without plasmid?

   _____

Summary of Activities

1. Isolate and characterize DNA from *Halobacterium*, a "salt-loving" bacterium.
2. Determine how heat and low pH affect the viscosity of DNA.
3. Transform a common bacterium into an ampicillin-resistant bacterium.

Questions for Further Thought and Study

1. How frequently do genetic transformations occur in nature? Justify your answer.
2. How could genetic transformations improve our quality of life? How could they decrease our quality of life?
3. Why is molecular biology often referred to as genetic engineering or biotechnology?

Doing Botany Yourself

Design and conduct an experiment to test the effects of acid pH on the integrity of isolated DNA. How do the results compare to the effects of basic solutions on DNA?

Writing to Learn Botany

Many people resist the use of genetic engineering to alter organisms. What are their arguments? Do you agree?

14

Plant Tissues

Objectives

By the end of this exercise you will be able to

1. describe the characteristics of parenchyma, collenchyma, sclerenchyma, epidermis, and vascular tissue
2. understand the structural variations exhibited by the cell types that form different tissues.

The structure of plants varies greatly among species; compare, for example, an oak tree with a cactus. However, these structural differences are usually quantitative rather than qualitative. That is, leaves, roots, and stems are made of the same types of cells and tissues; the major structural differences of organs such as stems and leaves result not from unique tissues but from these tissues being arranged differently and occurring in different proportions. All of these differences among plants represent different ways of achieving the same goals; namely, survival and reproduction.

PARENCHYMA, COLLENCHYMA, AND SCLERENCHYMA

Parenchyma, collenchyma, and sclerenchyma are types of ground tissue, a tissue that forms most of the primary body of a plant. These tissues have many functions, including nutrient storage, support, and basic metabolism such as photosynthesis. As you study individual cells, remember that the classification of plant cells as either parenchyma, collenchyma, or sclerenchyma is somewhat arbitrary; the characteristics of these tissue and cell types often overlap and intergrade.

Parenchyma cells are the most abundant and versatile cells in plants. These cells exhibit relatively few distinctive structural characteristics. In fact, any cell type not assignable to another functional or structural group is classified as parenchyma.

Parenchyma cells are living cells that are physiologically active in synthesis, transport, or storage of metabolic products. These cells can divide even at maturity and can differentiate into other cell types. Parenchyma cells often are polyhedral and accompanied by obvious intercellular spaces. Their walls are usually primary, although some parenchyma cells have secondary walls.

Parenchyma occurs throughout a plant. For example, pith, cortex, the fleshy tissue of fruits and seeds, the photosynthetic cells of leaves, and the rays of secondary xylem and phloem are composed of parenchyma cells. Parenchyma is also the basic ground tissue from which other more specialized cell types presumably have evolved.

Collenchyma consists of living, elongated cells that have unevenly thickened primary walls. Collenchyma cells are arranged in strands and occur only in the primary plant body, usually in elongating regions of shoots. In spite of their strength, collenchyma cells have plastic, or stretchable, primary walls. This feature enables collenchyma to elongate as the tissue it supports elongates. Collenchyma supports elongating organs such as young stems, leaf petioles, and flower and fruit stalks.

Collenchyma cells differentiate from parenchyma cells. The different types of collenchyma are recognized based on the pattern of wall thickening. The two most common types of wall thickenings are angular and lamellar. In angular collenchyma the cell corners are differentially thickened. Lamellar collenchyma results from thickenings on the inner and outer tangential walls (i.e., the walls parallel to the outside surface).

Sclerenchyma cells are rigid cells with thick, often lignified secondary walls. Like collenchyma, sclerenchyma functions in support. However, sclerenchyma cells cannot grow and often are dead at maturity. Thus, sclerenchyma supports nonelongating, mature regions of plants. Botanists recognize two groups of sclerenchyma cells: sclereids and fibers. **Sclereids** are small, irregularly shaped cells that usually occur singly or in small groups. **Fibers** are long, thin cells that typically occur in strands. You will study commercial fibers in a later exercise in this manual.

Procedure 14.1
Examine parenchyma

In this procedure you will examine a few of the many variations in parenchyma cells. During the remainder of the course, you'll have occasion to observe many of the remaining variations.

1. Review the appearance of a storage parenchyma cell by observing a *Helianthus* (sunflower) stem in cross section (x.s.) and longitudinal section (l.s.).

Question 1
a. Are intercellular spaces evident? _____

b. Are these cells meristematic? _____
   How do you know? _____
   _____

c. Are all parenchyma cells in the same tissue region of uniform size? _____

2. Observe the stellate parenchyma cells in the *Juncus* stem. Such tissue is common in the stems and roots of aquatic plants.

Question 2
a. How would you describe the shape of these cells?
   _____

b. How would you describe the intercellular spaces?
   _____
   What are their functions? _____
   _____

c. Why do we call this tissue "aerenchyma"? _____
   _____

Procedure 14.2
Examine collenchyma

1. Obtain a prepared slide of *Coleus* stem, x.s. and l.s. Notice that the stem is square in cross section. Collenchyma strands are in the four corners of the stem outside of the vascular bundles. Examine the

collenchyma cells using high power, noting the cell lumens surrounded by the irregularly thickened cell walls.

Question 3
a. Where are the thickenings located? _____
   _____
   What type of collenchyma is this? _____

b. Are intercellular spaces present? _____
   If so, where? _____
   _____

c. Are the cells uniform or do intermediate types exist adjacent to the parenchyma? _____
   _____

d. How does a collenchyma cell differ in appearance from a parenchyma cell? _____
   _____

2. Make a hand section of a fresh *Apium* (celery) petiole and stain it with safranin or crystal violet. Notice that the petiole is ribbed along the outside. In each rib, just beneath the epidermis, is a strand of collenchyma. When studying the collenchyma, don't confuse it with the vascular bundles located just inside the collenchyma strands.

3. Remove a strand by peeling away one of the ribs with your fingernail. Once you start the rib peel, you can strip off the entire strand along the length of the petiole. Test the strength of the strand by pulling it apart.

4. Now use a razor blade to cut a strip of parenchyma tissue from the inside of the petiole. Compare the strength of this tissue with that of collenchyma.

Question 4
a. Can you think of a function for the collenchyma strands of celery? _____
   _____

b. How does the strength of a strip of parenchyma tissue compare with that of collenchyma? _____
   _____

5. Study a prepared slide of a *Helianthus stem*, x.s. and l.s. Collenchyma cells are located just beneath the epidermis and extend all around the stem.

Question 5
a. Where are the irregular wall thickenings? _____
   _____

b. What type of collenchyma is this? _____

Procedure 14.3
Examine sclerenchyma

1. Stone cells, or brachysclereids, are small, roughly isodiametric sclereids. To observe a stone cell, mount a small amount of pulp from the fruit of a pear (*Pyrus*) on a slide. Add a drop of phloroglucinol and wait 10 to 15 minutes. Gently tap the coverslip to flatten the pulp. Now locate the thick-walled stone cells on the slide.

Question 6
a. What is the shape of a stone cell? _____

b. What characteristic of pear fruit do you think stone cells are responsible for? _____

2. Obtain a prepared slide of a *Hoya* stem. Observe the appearance of the stone cells in the pith, or center, of the stem.

Question 7
a. How do these cells differ from those in *Pyrus*? _____

_____

b. What might be the function of such a cell? _____

_____

3. Astrosclereids, or star cells, have branches diverging from a central body. Observe a prepared slide of a *Camellia* petiole.

Question 8
a. Are the sclereids distributed uniformly throughout the tissue? _____

4. Observe the fiber bundles in a cross section of *Vitis* wood. Remember that the fibers are small in cross section and have very thick walls.

Question 9
a. Where do fibers occur in this stem? _____

b. What purpose might they serve? _____

c. What characteristics enable you to identify fibers in sectioned material? _____

_____

EPIDERMIS

The epidermis is the dermal tissue that covers the primary plant body. It is the outermost tissue of the primary plant body, and it is replaced by the periderm (a tissue you'll learn about in a future exercise) during secondary growth. The epidermis has several functions, including (1) protection against injury (e.g., by pathogens) and desiccation, (2) gas exchange through specialized structures called stomata, (3) absorption (especially by the epidermis of the root), and (4) secretion.

Procedure 14.4
Examine the epidermis

1. Make an epidermal peel of a leaf of *Kalanchoë*. Your instructor will demonstrate how to do this. In brief, crease the leaf so that the surface to be peeled is on the inside. The fleshy tissue of the leaf should break, but the epidermis in the crease should not. Peel away the fleshy tissue from the intact epidermis. Mount the peel with the outside surface upward in a drop of 50% ethanol (to help eliminate air bubbles).

2. Examine the epidermal peel.

Question 10
a. Are the stomata arranged randomly or in a pattern?

_____

b. What is the shape of the cells that form the epidermal pore? _____

3. Examine epidermis in cross section by studying prepared slides of cross sections of leaves of *Ligustrum* (privet) and *Zea mays* (corn).

Question 11
a. How are the epidermal tissues of these plants different? _____

_____

How are they alike? _____

_____

Trichomes are unicellular and multicellular appendages of epidermal cells. Trichomes have a variety of structures and functions. In this lab you'll examine just one example (stinging hairs). In exercise 16, entitled "Roots," you'll examine another example of trichomes: root hairs.

Procedure 14.5
Examine trichomes

Examine a prepared slide of the stinging hairs of *Urtica* (stinging nettle). Sketch a stinging hair in the space below.

Question 12

How does the structure of a hair correlate with its function? _____

_____

VASCULAR TISSUES

The vascular tissues in plants are xylem and phloem. Xylem conducts water and dissolved minerals up the plant under a negative pressure. The conducting cells in xylem of gymnosperms and angiosperms are tracheids and vessel elements. Phloem conducts water and inorganic solutes (especially sugars) in all directions in a plant. The conducting cells in phloem of angiosperms are sieve tube members.

Procedure 14.6
Examine xylem and phloem

1. Examine a prepared slide of a longitudinal section of a stem of cucumber.
2. To locate the phloem, first find the xylem, which is stained red. In cucumber, you'll find the green-stained phloem on both sides of the xylem.
3. Now examine the vascular bundle in cross section. Find a sieve plate.

Question 13
a. What cellular features did you use to locate the sieve plate? _____
b. What cellular components are visible in sieve tube members? _____

4. Sieve tube members are often associated with companion cells. Locate a companion cell.

Question 14
a. What is the function of a companion cell? _____

b. How is the structure of a companion cell different from that of a sieve tube member? _____

_____

5. Examine a prepared slide of a cross section of a stem of corn (*Zea mays*). Locate and sketch a vascular bundle.

Question 15
a. Where in the bundle is the phloem? _____

_____

6. Examine a prepared slide of a cross section of pine (*Pinus*) wood. Virtually every cell that you see is a tracheid.

Question 16
a. How is the structure of tracheids adapted for transport in the xylem? _____

_____

b. Aside from transport of water and dissolved solutes, what other function is performed by tracheids?

_____

7. Examine a prepared slide of a cross section of oak (*Quercus*) wood.

Question 17
a. How does the secondary xylem of oak differ from that of pine? _____

_____

b. What are the large cells in the xylem? _____

_____

c. What is their function? _____

Summary of Activities

1. Examine parenchyma cells in *Helianthus* stem.
2. Examine aerenchyma cells in *Juncus* stem.
3. Make a hand section of a petiole of *Apium*.
4. Examine collenchyma in *Helianthus* stem.
5. Examine stone cells in *Pyrus* and *Hoya*.
6. Examine astrosclereids in *Camellia*.
7. Examine fibers in stems of *Vitis*.
8. Examine an intact epidermis from *Kalanchoë*.
9. Examine trichomes in *Urtica*.
10. Examine xylem and phloem in stems of cucumber, oak, and pine.

Questions for Further Thought and Study

1. How would you distinguish parenchyma from collenchyma? collenchyma from sclerenchyma?
2. How is the epidermis adapted for its protective function?
3. How does the structure of a vessel element differ from that of a sieve tube member? What is the functional significance of these differences?

Doing Botany Yourself

Examine cross sections of stems of some aquatic plants. What features do they have in common? Of what importance are these features to these specialized plants?

Writing to Learn Botany

Which is more versatile: parenchyma, collenchyma, or sclerenchyma? Why?

Commercial Fibers

Objectives

By the end of this exercise you will be able to
1. describe the structure of materials made from plant fibers
2. describe the economic importance of fibers.

The term "fiber" has two meanings to many botanists. The more restricted meaning refers to a long, tapered, thick-walled sclerenchyma cell. However, the term also is applied to plant tissues used by the textile and paper industries. To eliminate confusion, we refer to the latter type of fibers as "commercial fibers."

Commercial fibers have several useful features. For example, commercial fibers typically are strong due to the presence of secondary walls. They also are somewhat pliable, with the exception of hard fibers that have a high lignin content. Commercial fibers also tend to be elongate, thereby allowing the fibers to be interwoven. Many plants contain tissues that have these characteristics and thus can be utilized as commercial fibers. Such plants are exploited heavily, particularly by people in nonindustrialized cultures. In this exercise, we'll consider only those fibers commonly encountered in our own society.

The textile industry uses commercial fibers to make products such as cloth and rope. Cotton fiber is obtained from **Gossypium** and consists of elongate epidermal trichomes on the seed coat. Flax fibers are used to make linen. These fibers consist of bundles of true sclerenchyma fibers from **Linum** stems. Ramie fiber consists of sclerenchyma fiber bundles from **Boehmeria** and is used to make cloth. Jute fiber consists of sclerenchyma bundles from **Corchorus;** jute fibers are used in products such as burlap, mats, and rope. Fibers of **Cannabis sativa,** or hemp, consist of sclerenchyma bundles. Hemp originally was introduced and cultivated in the central United States for use in making rope. Entire vascular bundles from leaves of **Agave** form sisal, which is used in the manufacture of rope.

Commercial fibers are also used in paper products. All paper and paper products are made from wood. The uniformity of the wood fibers is important in making high-quality paper. Conifer wood is used most often because of its homogeneous content of tracheids. Hardwoods also are used to make paper, but these woods contain a greater assortment of cell types. To make paper more pliable, the wood is treated to remove much of the lignin.

Procedure 15.1
Examine textile fibers

Cotton
1. Examine a prepared slide of an ovary of *Gossypium* (cotton) with your low-power objective. You will see a four-chambered fruit. Within each of the chambers, or locules, you will see three to five developing seeds that are sectioned in various planes. Examine the epidermal structure of one of these seeds and observe the early development of epidermal trichomes. It is these trichomes on the seed coat that constitute cotton fiber.

2. A cotton boll is a dry fruit that tears apart to reveal the four seed chambers. At maturity each chamber is filled with seeds, each of which has its own sheath of trichomes. Obtain a mature cotton boll and relate its structure to what you observed on the slide. Also mount a trichome or two in a drop of water on a slide, stain with safranin, and observe.

3. Identify the fruit (ovary) wall, the four seed chambers, the seeds, and the trichomes of the cotton boll.

Question 1

a. Does the entire surface or only a portion of the seed coat form trichomes? _____

b. Are the trichomes attached to the seeds? _____ _____

c. Approximately how long is a mature trichome? _____

d. Is a secondary wall present? _____

   If so, is it lignified? _____

e. What effect does lignin have on the quality of fibers? _____

f. What function might these trichomes serve in nature? _____

Flax

1. To study the anatomy of the sclerenchyma fibers of flax, observe prepared slides of *Linum* stem, x.s. and l.s.

2. Study linen textile by preparing two slides of macerated linen cloth. Stain one preparation with phloroglucinol and the other with safranin. After several minutes observe the fibers.

Question 2

a. Where in the stem are the sclerenchyma fibers located? _____

b. Are these fibers relatively short or long? _____ _____

c. Why are these fibers called soft fibers? _____ _____

d. Have the fibers been processed in any observable way? _____

   How might you know? _____ _____

Sisal

Although true sisal is obtained from one particular species of *Agave*, most species in this genus yield fiber strands that consist of entire vascular bundles. Tear out a vascular bundle from the fresh leaves of *Agave*. Make a fresh cross section of a bundle; stain and observe the preparation.

Question 3

a. What tissues are responsible for the strength of the fibers? _____

b. For what is sisal used? _____

   How does its structure aid in this use? _____ _____

Hemp

Observe the prepared slides of *Cannabis sativa* stem, x.s.

Question 4

a. Are any fibers present? _____

b. If so, in what tissue are they located? _____ _____

Procedure 15.2

Examine paper

Prepare macerated preparations of at least two of the papers provided in lab. Stain separate preparations with phloroglucinol and safranin and observe after several minutes.

Question 5

a. What kinds of cells occur in the various papers examined? _____

b. How are the cell types correlated with the quality of paper? _____

c. What is the lignin content of the paper fibers? _____

d. What other fibers can be used to make paper? _____ _____

Summary of Activities

1. Examine textile fibers of cotton, flax, sisal, and hemp.
2. Examine fibers in paper products.

Questions for Further Thought and Study

1. What is the significance of fibers in plants?
2. What features of fibers make them commercially useful?

16

Roots

Objectives

By the end of this exercise you will be able to
1. describe the functions of roots
2. describe the structural and functional differences between tap and fibrous root systems
3. describe the structure of roots
4. describe the origin of secondary and adventitious roots.

During seed germination a **radicle** or young **primary root** emerges from the seed and grows down. The primary root soon produces numerous **secondary roots** and forms a root system, whose major functions include absorbing water and minerals, anchoring the plant, and storing food. Root systems have different shapes. For example, a **tap root system** has a large main root and smaller secondary roots branching from it (as in the carrot). In a **fibrous root system,** the primary and secondary roots are similar in size (as in many grasses). Examine the displays of tap and fibrous root systems available in the lab.

Question 1
a. Which root system—tap or fibrous—absorbs water more effectively? _____

Why? _____

b. Which root system is a more effective anchor? _____

_____

Why? _____

_____

c. Which root system is more effective for storage?

_____

Why? _____

_____

d. What is the adaptive significance of the fact that not all roots grow straight down? _____

_____

ROOT GROWTH

Primary growth and all primary tissues are formed by **apical meristems,** which occur at the tips of roots and stems. A meristem is a localized area of cellular division. Primary growth (growth in length) produces herbaceous (non-woody) tissue. **Secondary growth** refers to growth in girth resulting from nonapical meristems. Secondary growth is discussed in exercise 17, "Stems".

Procedure 16.1
Examine the root apex

1. Examine the root tips of 2-day-old seedlings of radish (*Raphanus*) and corn (*Zea*) with your dissecting microscope.
2. Referring to figure 16.1, identify the root cap, root apical meristem, zone of elongation, and zone of maturation.

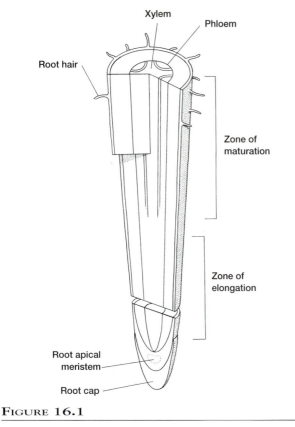

Xylem

Phloem

Root hair

Zone of
maturation

Zone of
elongation

Root apical
meristem

Root cap

FIGURE 16.1

Root tip showing root cap, root apical meristem, zone of elongation, and zone of maturation.

A thimble-shaped **root cap** covers the root tip and protects the root apical meristem. The root cap protects the root by secreting mucigel and sloughing cells as the root grows through the soil. The **root apical meristem** occurs behind the root cap and produces the new cells for primary growth. These cells elongate in the **zone of elongation,** which is 2 to 6 mm behind the root tip. This elongation produces primary growth. The **zone of maturation** is an area in which cell division is infrequent and cells are maturing metabolically and morphologically to assume their functions in the mature root.

Question 2

a. In which area of a root tip are cells the largest?

_____

In which area are they the smallest? _____

b. Aside from their size, do all cells in the root tip

appear similar? _____

c. What do you conclude from this observation?

_____

At the edge of the root cap are peripheral cells, which secrete mucigel. Peripheral cells surround the columella tissue, which is located in the center of the root

cap. Columella tissue is suspected by some botanists to be the site of gravity perception by roots.

Question 3

a. What are the distinguishing features of peripheral

cells? _____

b. What are the distinguishing features of columella

cells? _____

Procedure 16.2
Examine root hairs

1. Reexamine the 2-day-old radish seedling with your dissecting microscope.

2. Note the **root hairs** in the zone of maturation. Root hairs are outgrowths of epidermal cells and are short-lived. Root hairs increase the surface area of the root.

Question 4

a. Why do you think root hairs occur only in the zone

of maturation? _____

_____

b. Approximately how many root hairs are on your

seedling? _____

c. What are the functions of root hairs? _____

_____

PRIMARY TISSUES OF THE ROOT

The root apical meristem produces cells that differentiate into primary tissues of the root. The outer layer of cells is the **epidermis.** Just inside the epidermis is the **cortex,** whose cells contain numerous **amyloplasts** (starch-containing plastids). The inner layer of the cortex is the **endodermis,** which regulates water flow to the vascular tissue in the center of the root. Immediately inside the endodermis is the **pericycle,** which can become meristematic and produce **secondary** (i.e., **lateral**) **roots.**

Procedure 16.3
Examine primary tissues of a root

Monocot Root

1. Examine a prepared slide of a cross section of a root of *Smilax,* a monocot. Locate the cortex, epidermis, pericycle, and stele. Note the different structures of these tissues.

2. Examine the endodermis closely. In *Smilax,* the suberized walls of the endodermis will be obvious. Suberin deposits on the cell walls make them impervious to water.

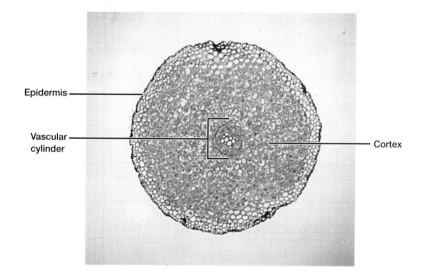

FIGURE 16.2

Cross section of root of buttercup (*Ranunculus* sp.). At this magnification, the most obvious tissues are the epidermis, cortex, and the vascular tissues of the vascular cylinder.

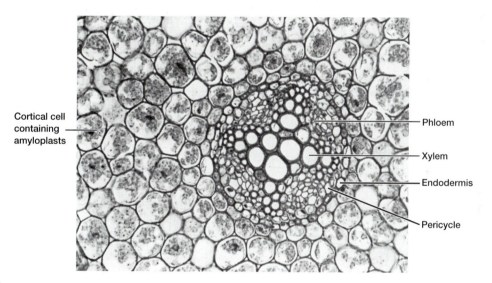

FIGURE 16.3

A cross section through the region of maturation of a root of a buttercup (*Ranunculus* sp.), a dicot.

Question 5

a. Are all of the cells of the endodermis alike? _____

b. Is there any evidence of suberization in cells other than the endodermis? _____
    What does this tell you about the developmental control of suberization? _____

c. Are intercellular spaces prominent in the endodermis as compared to the cortex? _____
    What is the functional significance of this? _____
    _____

Dicot Root

1. Examine a prepared slide of a cross section of a root of buttercup (*Ranunculus*), a dicot (figs. 16.2 and 16.3).

2. Examine a prepared slide labeled "lateral root origin."

3. Locate the epidermis, cortex, pericycle, and newly formed lateral root.

Question 6

a. What do you suppose is the primary function of the cortex? _____
    _____

b. Do secondary roots have root caps? _____

c. What happens to tissues of the main root (i.e., cortex, endodermis, and epidermis) as the lateral root grows? _____
_____

d. Do secondary roots arise inside the primary root or on its surface? _____

e. How do the epidermis and cortex of *Ranunculus* compare with those of the *Smilax* root that you just studied? _____
_____

In the center of the buttercup root is the **vascular** (fluid-conducting) **cylinder** composed of xylem and phloem (fig. 16.3). **Xylem** transports water and minerals, while **phloem** transports most organic compounds in the plant. Water-conducting cells in the xylem of angiosperms are called tracheids and vessels, and they are dead and hollow at maturity. **Vessels** are stacks of cylindrical cells with thin or completely open end-walls. Water moves through vessels in straight, open tubes. These tubes are usually stained red in slide preparations of buttercup roots. **Tracheids** are long, spindle-shaped conducting cells; tracheids are the only conducting cells in wood of gymnosperms such as pine. Tracheids and vessels have thin areas called **pits,** where the cell walls of adjacent cells overlap (fig. 16.4). Water meanders through these pits from one cell to the next.

Conducting cells in phloem are called **sieve cells** and **sieve tube members,** and they are alive at maturity. **Phloem** consists of small, thin-walled cells arranged in bundles that alternate with the poles of xylem. Sieve tube members are usually stained green.

Procedure 16.4
Examine carrot root

1. Prepare two thin cross sections of a carrot root.
2. Stain one slice with iodine (a stain for starch) and examine it with your microscope.

3. Stain the other section of carrot root with phloroglucinol. Phloroglucinol stains **lignin,** a molecule that strengthens xylary cell walls.

Question 7
a. Where is the starch located in carrot root? _____
_____

b. What can you conclude from this observation?
_____

c. How does the arrangement of xylem in a carrot root differ from that in a buttercup root? _____
_____

Adventitious Roots
Adventitious roots are those that arise from the shoot system, either regularly (as in prop roots of corn) or induced (as in roots on cuttings).

Procedure 16.5
Examine adventitious roots

1. Examine the adventitious roots on the cuttings on display.
2. Examine a prepared slide of adventitious roots of tomato.

Question 8
a. From what tissue or tissues do adventitious roots arise?
_____

Summary of Activities

1. Examine tap and fibrous root systems.
2. Examine root tips of radish and corn.
3. Examine a prepared slide of the root tip of corn or onion.
4. Examine prepared slides of roots of *Ranunculus* and *Smilax.*
5. Stain and examine carrot root.
6. Examine prepared slides showing the origin of lateral and adventitious roots.

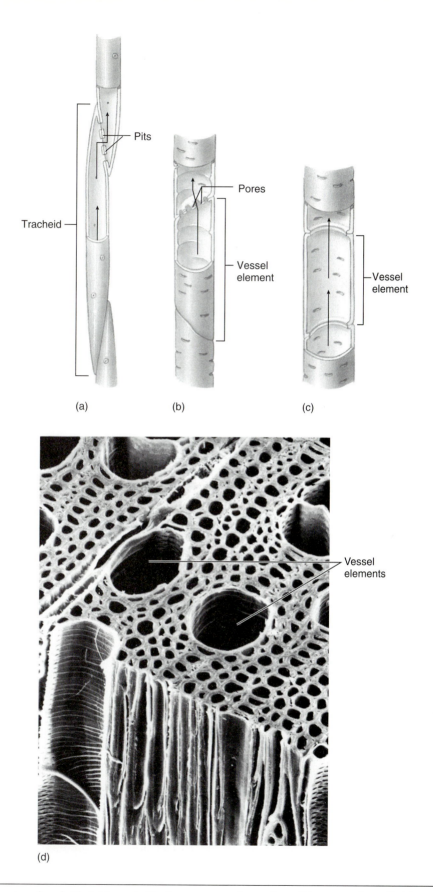

(a)

(b)

(c)

(d)

FIGURE 16.4

Comparison of vessel elements and tracheids. (*a*) In tracheids, the water passes from cell to cell by means of pits. (*b,c*) In vessel elements, water moves by way of pores, which may be simple or interrupted by bars. (*d*) The large openings shown in this scanning electron micrograph of the wood of a red maple (*Acer rubrum*) are vessel elements (×350).

Questions for Further Thought and Study

1. Lateral roots are most abundant in the region just behind the zone of root hairs. What is the functional significance of this?

2. An exception to the previous statement occurs in floating aquatic plants, where the lateral roots form near the root tip. How do you explain this?

3. Do roots have a prominent cuticle? What is the functional significance of this?

Doing Botany Yourself

Roots grow down. Design and conduct an experiment using corn seeds to demonstrate if this is a negative response to light or a positive response to gravity.

Writing to Learn Botany

What is the general functional significance of lateral roots? root hairs? root caps? In some plants, one or more of these structures are extremely reduced. What kinds of plants would you expect these to be?

Stems

17

Objectives

By the end of this exercise you will be able to
1. understand the structure and function of stems
2. describe the external features of woody stems
3. describe the primary structure of monocot stems
4. describe the secondary growth of stems.

Stems are often the most conspicuous organs of plants, and they function for support and transport of water and solutes (see plate VI, fig. 11). Some stems (e.g., those of cacti) also photosynthesize and store food.

THE SHOOT APEX

Procedure 17.1
Examine the shoot apex

1. Examine a living *Coleus* plant and note the arrangement of leaves on the stem.
2. Keeping this observation in mind, examine a prepared slide of a longitudinal section of a shoot tip of *Coleus* (fig. 17.1). Note that the dome-shaped **shoot apical meristem** is not covered by a cap as is a root. The shoot apical meristem produces young leaves, or **leaf primordia,** that attach to the stem at a **node.** An **axillary bud** between the young leaf and the stem forms a branch or flower.

Question 1
a. How does the absence of a cap at a shoot apex differ from the apical meristem of a root? _____
_____
b. What is the significance of this difference? _____
_____
c. Are all cells in the shoot apex the same size? _____
What do you conclude from this observation? _____
_____

Procedure 17.2
External features of a mature woody stem

1. Examine the features of a dormant twig (fig. 17.2). A **terminal bud** containing the apical meristem is at the stem tip and is surrounded by **bud scales. Leaf scars** from shed leaves occur at regularly spaced **nodes** along the stem length. The portion of the stem between the nodes is called the **internode. Vascular bundle scars** may be visible within the leaf scars. Axillary buds protrude from the stem just distal to each leaf scar.
2. Search for clusters of **bud scale scars.** The distance between clusters or from a cluster to the terminal bud indicates the length of yearly growth.

Procedure 17.3
Examine primary tissues of dicot stems

1. Examine a prepared slide of a cross section of a sunflower (*Helianthus*) stem (fig. 17.3). An **epidermis** covers the stem. The epidermis is coated with a waxy, waterproof substance called **cutin.** Below the epidermis is the **cortex,** which stores food. The **pith** in the center of the stem also stores food. In sunflower stems, the cortex is not uniform. Rather, the three to four cell layers of the cortex just

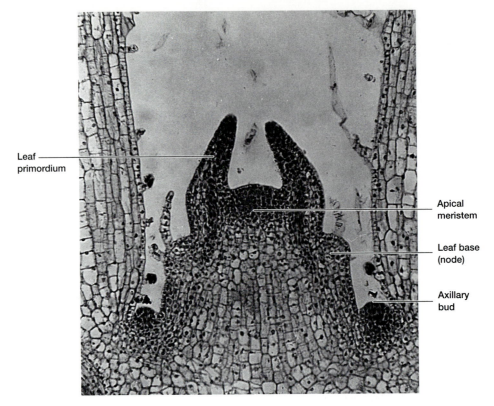

Leaf primordium

Apical meristem

Leaf base (node)

Axillary bud

FIGURE 17.1

Coleus shoot tip. The apical meristem is the site of rapid cell division and primary growth. Young leaves are produced at the tip of the shoot. Shoot tips, unlike root tips, are not covered by a cap.

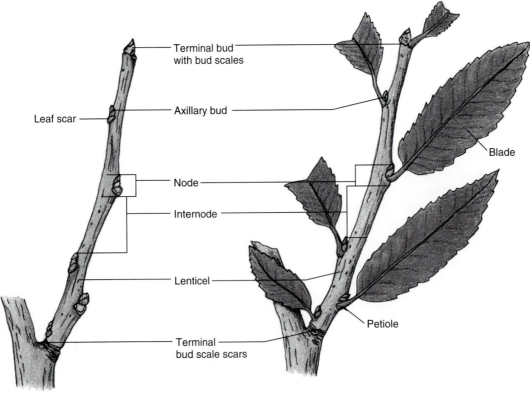

Terminal bud with bud scales

Axillary bud

Leaf scar

Blade

Node

Internode

Lenticel

Petiole

Terminal bud scale scars

FIGURE 17.2

A woody twig. (*a*) The twig in its dormant winter condition. (*b*) The same twig as it appeared the summer before.

17-2

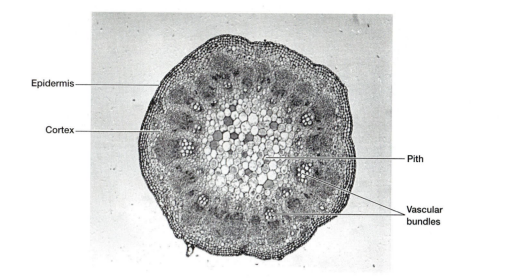

FIGURE 17.3

Sunflower stem, cross section.

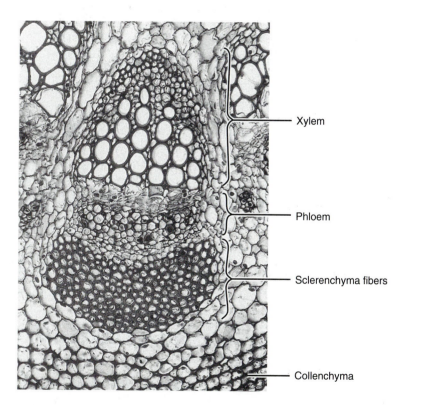

FIGURE 17.4

Cross-sectional view of vascular tissue of a sunflower (*Helianthus annuus*) stem , ×485. The sclerenchyma fibers support the stem and protect the vascular tissue. Note that the walls of fibers are much thicker than those of adjacent cells.

below the epidermis are smaller, rectangular cells with thickened cell walls. These are **collenchyma** cells; the cells support elongating regions of the plant.

2. In stems, xylem and phloem are arranged in bundles (fig. 17.4). Locate one of these bundles on the slide you are examining. Sketch what you see.

Question 2

a. What is the significance of a coating of cutin?

_____

b. How does the arrangement of xylem and phloem in stems differ from that in roots? _____

_____

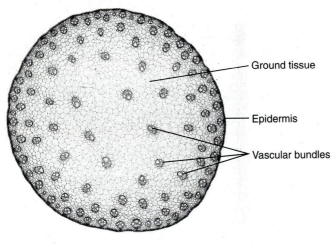

Ground tissue

Epidermis

Vascular bundles

FIGURE 17.5

Cross section of a stem of corn (*Zea mays*), a monocot, ×5. Unlike in dicots such as the sunflower, bundles of vascular tissue in monocots occur throughout the ground tissue. The stem is surrounded by an epidermis.

c.   In stems, which vascular tissue is on the inside of the

bundle? _____

d.   Which tissue is on the outside of a vascular bundle?

_____

The large, darkly stained, thick-walled cells in figure 17.4 are **sclerenchyma fibers,** which function in support. Sclerenchyma fibers from some plants are used to make linen, rope, and burlap (see exercise 15, "Commercial Fibers").

The ring of vascular bundles in sunflower stems is typical of **dicots,** which are flowering plants with two **cotyledons** (seed leaves).

Procedure 17.4
Examine primary tissues of monocot stems

1.   Examine a prepared slide of a cross section of a corn stem (fig. 17.5). Corn is a **monocot,** a flowering plant with only one cotyledon.

2.   In the space below, sketch a cross section of a sunflower stem and a corn stem. Note the distribution of the vascular bundles.

Question 3
a.   How does the arrangement of vascular bundles differ in stems of monocots as compared to dicots?

SECONDARY GROWTH OF STEMS

Between the xylem and phloem in each vascular bundle in dicot stems is a meristematic tissue called the **vascular cambium.** The vascular cambium is a secondary meristem that produces secondary growth (i.e., growth in girth). The vascular cambium is cylindrical, and it produces secondary xylem to its inside and secondary phloem to its outside.

Question 4
a.   How is secondary growth different from primary

growth? _____

Procedure 17.5
Secondary growth of stems

1.   Examine a cross section of a 1-year-old basswood (*Tilia*) stem. The vascular cambium is a narrow band of cells between the xylem and phloem (fig. 17.6).

2.   Examine the structure of 2- and 3-year-old stems of basswood. Compare them with a 1-year-old stem. Note that the secondary xylem of older stems consists of concentric annual rings made of alternating layers of large and small cells. The large cells are formed in the spring, while the smaller cells are produced in summer. In the space below, draw 1-, 2-, and 3-year-old cross sections of the stems.

Question 5
a.   How do you account for this seasonal production of

different-sized cells? _____

_____

b.   What is the common name for secondary xylem?

_____

c.   What is "grain" in wood? _____

_____

d.   Aside from conducting water and minerals, what is another important function of secondary xylem?

_____

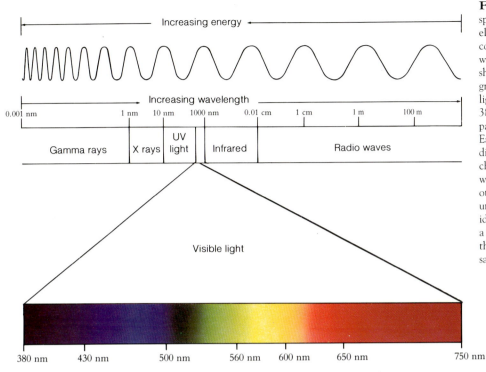

FIGURE 1 Electromagnetic spectrum. Light is a form of electromagnetic energy and is conveniently thought of as a wave with a repetitive wavelength. The shorter the wavelength of light, the greater the energy of the light. Visible light includes wavelengths between 380 and 750 nm, which is only a small part of the electromagnetic spectrum. Each wavelength of light represents a different color. Atoms, molecules, and chemical bonds absorb some wavelengths of light more than others. Because each chemical has a unique absorption spectrum, we can identify and quantify chemicals using a spectrophotometer that measures the light absorbed by an unknown sample (see Appendix B).

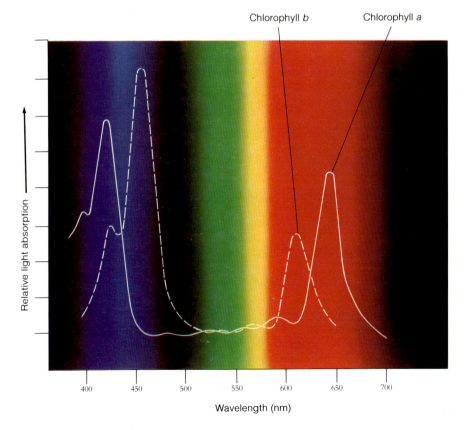

FIGURE 2 Absorption spectra for chlorophyll *a* and chlorophyll *b*. Chlorophyll *a*, along with other accessory pigments, occurs in all photosynthetic plants. Instructions for measuring the absorption spectrum of a pigmented plant extract are provided in Appendix B.

PLATE   I

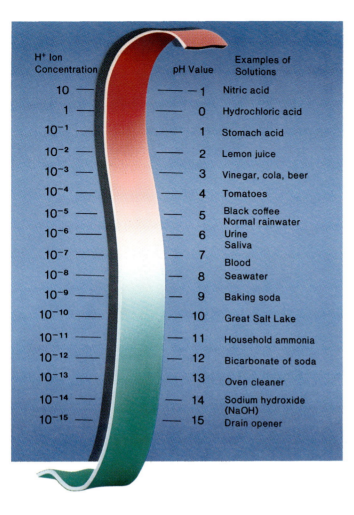

H+ Ion Concentration	pH Value	Examples of Solutions
10	-1	Nitric acid
1	0	Hydrochloric acid
10^{-1}	1	Stomach acid
10^{-2}	2	Lemon juice
10^{-3}	3	Vinegar, cola, beer
10^{-4}	4	Tomatoes
10^{-5}	5	Black coffee, Normal rainwater
10^{-6}	6	Urine, Saliva
10^{-7}	7	Blood
10^{-8}	8	Seawater
10^{-9}	9	Baking soda
10^{-10}	10	Great Salt Lake
10^{-11}	11	Household ammonia
10^{-12}	12	Bicarbonate of soda
10^{-13}	13	Oven cleaner
10^{-14}	14	Sodium hydroxide (NaOH)
10^{-15}	15	Drain opener

FIGURE 3   The pH scale. A fluid is assigned a value according to the number of hydrogen ions present in a liter of that fluid. The scale is logarithmic, so that a change of only 1 means a tenfold change in the concentration of hydrogen ions; thus lemon juice is 100 times more acidic than tomato juice, and seawater is 10 times more basic than pure water.

FIGURE 4   DNA. A molecule of DNA is not a straight chain, but rather a graceful double helix rising like a circular staircase. Hydrogen bonds between organic bases bind the two chains of a DNA duplex to each other. These precise bonds and resulting base pairing give DNA its characteristically spiral shape. The chemical nature of DNA is described in Exercise 2.

Hydrogen bond

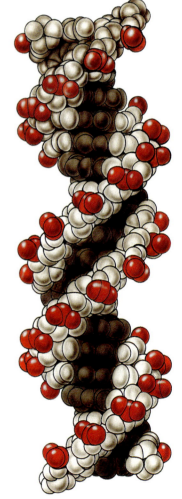

DNA double helix

PLATE   II

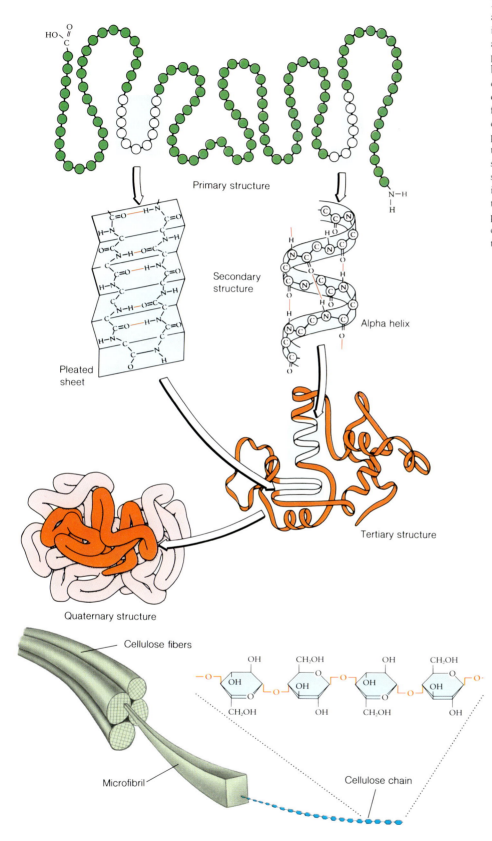

Primary structure

Secondary structure

Pleated sheet

Alpha helix

Tertiary structure

Quaternary structure

FIGURE 5 Proteins are made of amino acids and are biologically important macromolecules. The amino acid sequence of a protein is called its primary structure. Hydrogen bonds between nearby amino acids produce coils called alpha-helices and foldbacks called pleated sheets. These coils and foldbacks are the secondary structure of the protein. Regularly coiled and pleated proteins further flex and curve the protein into a more irregular shape supported loosely by hydrogen bonds, sulfide bonds, and hydrophobic interactions. This configuration is the tertiary structure of the protein. Many proteins aggregate in clusters like the one illustrated here. Such clustering is the quaternary structure of the protein.

Cellulose fibers

Microfibril

Cellulose chain

FIGURE 6 Cellulose is a polymer of glucose molecules and is the most abundant organic compound on earth. Free hydroxyl (OH⁻) groups of the glucose molecules form hydrogen bonds between adjacent cellulose molecules to form cohesive microfibrils. Microfibrils align to form strong cellulose fibers that resist metabolic breakdown. Because humans cannot hydrolyze the bonds between glucose molecules of cellulose, cellulose is indigestible and its energy is unavailable. Cellulose passes through the human digestive tract as bulk fiber.

PLATE III

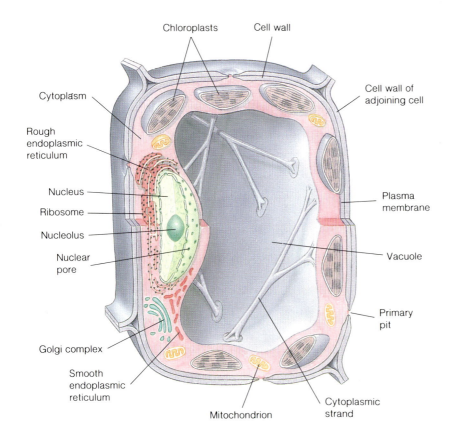

Chloroplasts

Cell wall

Cytoplasm

Rough
endoplasmic
reticulum

Nucleus

Ribosome

Nucleolus

Nuclear
pore

Golgi complex

Smooth
endoplasmic
reticulum

Mitochondrion

Cell wall of
adjoining cell

Plasma
membrane

Vacuole

Primary
pit

Cytoplasmic
strand

FIGURE 7   Plant cell. This
illustration shows relative proportions
of the different parts of a plant cell.
Most mature plant cells contain large
central vacuoles which occupy most
of the volume of the cell. Cytoplasm
occupies a thin layer between the
vacuole and plasma membrane and
contains all of the cell's mitochondria
and other organelles. Plant cells are
the subject of Exercise 4.

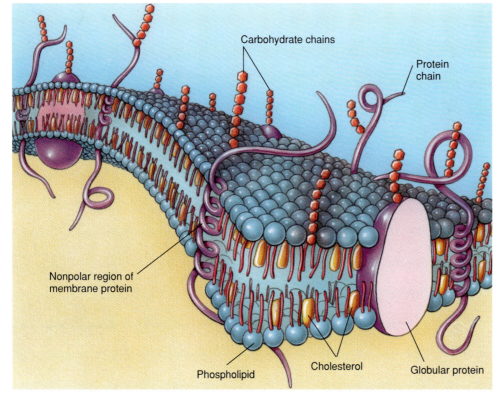

Carbohydrate chains

Protein
chain

Nonpolar region of
membrane protein

Phospholipid

Cholesterol

Globular protein

FIGURE 8   Cells have
complex surfaces. A variety of
proteins protrude through the
plasma membrane, nonpolar
regions of the proteins serving to
tether them to the membrane's
nonpolar interior. Carbohydrate
chains (strings of sugar molecules)
are often bound to these proteins
and to lipids in the membrane
itself as well. These chains serve as
distinctive identification tags that
are unique to particular types of
cells.

PLATE   IV

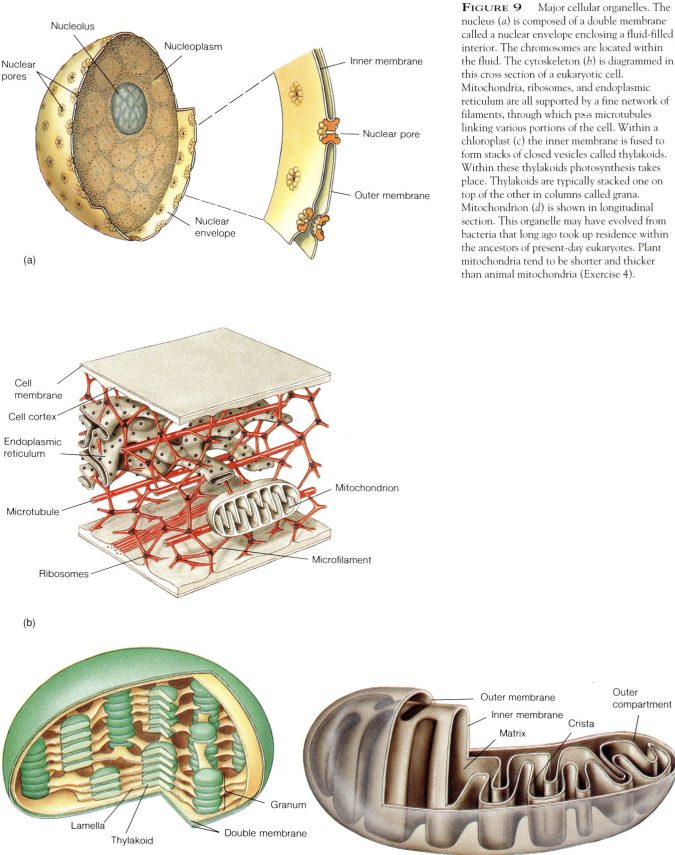

FIGURE 9 Major cellular organelles. The nucleus (*a*) is composed of a double membrane called a nuclear envelope enclosing a fluid-filled interior. The chromosomes are located within the fluid. The cytoskeleton (*b*) is diagrammed in this cross section of a eukaryotic cell. Mitochondria, ribosomes, and endoplasmic reticulum are all supported by a fine network of filaments, through which pass microtubules linking various portions of the cell. Within a chloroplast (*c*) the inner membrane is fused to form stacks of closed vesicles called thylakoids. Within these thylakoids photosynthesis takes place. Thylakoids are typically stacked one on top of the other in columns called grana. Mitochondrion (*d*) is shown in longitudinal section. This organelle may have evolved from bacteria that long ago took up residence within the ancestors of present-day eukaryotes. Plant mitochondria tend to be shorter and thicker than animal mitochondria (Exercise 4).

(a)

Nucleolus
Nucleoplasm
Nuclear pores
Inner membrane
Nuclear pore
Outer membrane
Nuclear envelope

(b)

Cell membrane
Cell cortex
Endoplasmic reticulum
Microtubule
Ribosomes
Mitochondrion
Microfilament

(c)

Lamella
Thylakoid
Granum
Double membrane

(d)

Outer membrane
Inner membrane
Matrix
Crista
Outer compartment

PLATE V

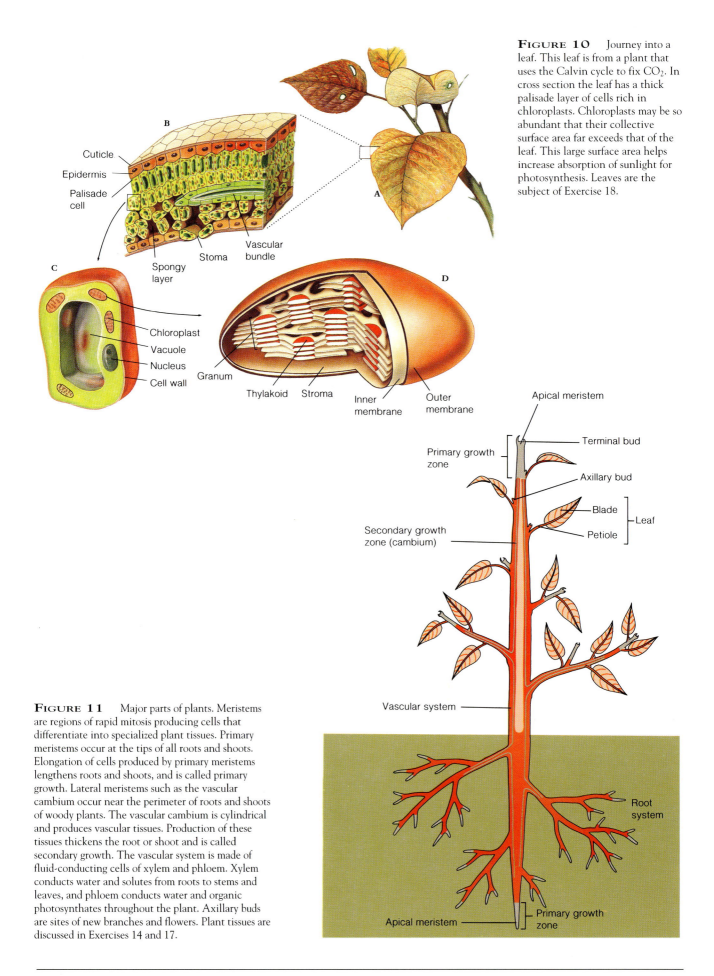

FIGURE 10 Journey into a leaf. This leaf is from a plant that uses the Calvin cycle to fix CO_2. In cross section the leaf has a thick palisade layer of cells rich in chloroplasts. Chloroplasts may be so abundant that their collective surface area far exceeds that of the leaf. This large surface area helps increase absorption of sunlight for photosynthesis. Leaves are the subject of Exercise 18.

B

Cuticle
Epidermis
Palisade cell
Stoma
Spongy layer
Vascular bundle

A

C

Chloroplast
Vacuole
Nucleus
Cell wall

D

Granum
Thylakoid
Stroma
Inner membrane
Outer membrane

Apical meristem
Terminal bud
Primary growth zone
Axillary bud
Blade
Petiole
Leaf
Secondary growth zone (cambium)
Vascular system
Root system
Apical meristem
Primary growth zone

FIGURE 11 Major parts of plants. Meristems are regions of rapid mitosis producing cells that differentiate into specialized plant tissues. Primary meristems occur at the tips of all roots and shoots. Elongation of cells produced by primary meristems lengthens roots and shoots, and is called primary growth. Lateral meristems such as the vascular cambium occur near the perimeter of roots and shoots of woody plants. The vascular cambium is cylindrical and produces vascular tissues. Production of these tissues thickens the root or shoot and is called secondary growth. The vascular system is made of fluid-conducting cells of xylem and phloem. Xylem conducts water and solutes from roots to stems and leaves, and phloem conducts water and organic photosynthates throughout the plant. Axillary buds are sites of new branches and flowers. Plant tissues are discussed in Exercises 14 and 17.

PLATE VI

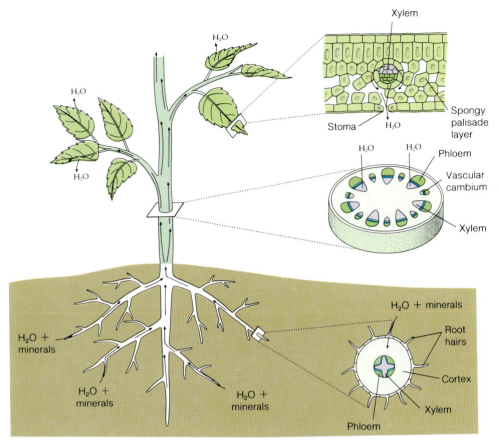

Xylem

Stoma

Spongy
palisade
layer

H_2O H_2O Phloem

Vascular
cambium

Xylem

H_2O + minerals

Root
hairs

Cortex

Xylem

Phloem

H_2O +
minerals

H_2O +
minerals

H_2O +
minerals

FIGURE 12 Flow of materials in a plant. Water with dissolved minerals enters the plant through root hairs and moves upward through the vascular system. Most of this water is eventually lost through the leaves during transpiration. Dissolved minerals are used by plant cells for metabolic chemical reactions. Carbohydrates made during photosynthesis move throughout the plant in the phloem. These carbohydrates are either stored or used as an energy source for metabolism and growth (Exercise 22).

FIGURE 13 (*a*) Two guard cells form each stoma on the surface of a leaf. When the concentration of solutes is high in the guard cells, water diffuses into the cells. This influx of water causes the cells to swell and bow apart, thereby forming a pore and opening the stoma. Water evaporates through these stomatal pores during transpiration. (*b*) When the solute pressure is low in the guard cells, the guard cells become limp and the stomatal pore closes. A variety of environmental factors that cause stomata to open and close are described in Exercise 22.

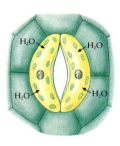

(a) Guard cells open

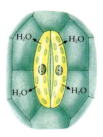

(b) Guard cells closed

PLATE VII

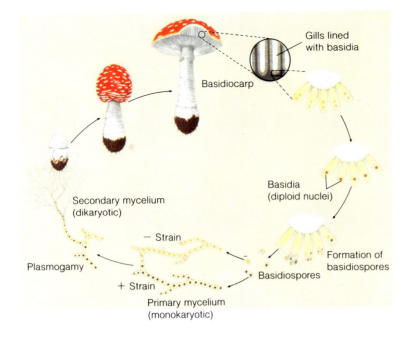

FIGURE 14 Life cycle of a basidiomycete (division Basidiomycota). The underground body of basidiomycetes, including mushrooms, is a mycelium of cellular filaments. In primary mycelia, each cell has only one nucleus. Secondary mycelia are formed by the fusion of primary mycelia and have two nuclei within each cell, one derived from each primary mycelium. Secondary mycelia ultimately become massed and interwoven to form conspicuous basidiocarps such as the familiar mushroom. Fusion and meiosis occur within basidia on gills of the basidiocarp. The resulting haploid basidiospores are released to continue the sexual reproductive cycle. A basidiospore germinates and produces the primary mycelium of a new organism (Exercise 25).

FIGURE 15 Fern life cycle. Small, heartshaped gametophytes grow on moist ground or similar places. Archegonia and antheridia on the surface of these haploid gametophytes contain developing eggs and sperm, respectively. Flagellated sperm are released from antheridia and swim to the mouth of the archegonium where they enter and fertilize the single egg. Fusion of egg and sperm forms a zygote, the first cell of the diploid sporophyte. The zygote begins to grow as a sporophyte within the archegonium. Eventually, the sporophyte becomes much larger than the gametophyte and becomes what we recognize as the fern plant. On its leaves form clusters of sporangia, within which meiosis produces haploid spores. Release of these spores, which is explosive in many ferns, and their germination produces new gametophytes (Exercise 28).

PLATE VIII

FIGURE 16 Sunflowers get their common name because their flowers track the sun across the sky, much as radiotelescopes track satellites or stars. This solar tracking, or heliotropism, by sunflowers and other plants (e.g., cotton, alfalfa, beans) helps the plants regulate the amount of light that they absorb. Desert plants such as the "compass plant" use heliotropism to orient their leaves parallel to the sun's rays, thereby minimizing the amount of light that they absorb and preventing overheating and desiccation. Other plants orient their leaves perpendicular to the sun's rays to maximize the amount of light absorbed for photosynthesis.

FIGURE 17 The giant leaves of this Amazon water lily contain many air spaces that make the leaves buoyant. These leaves can support a small child without sinking. This buoyancy, combined with leaves' large surface area, increases the amount of light absorbed for photosynthesis.

FIGURE 18 Bromeliads such as these often grow as epiphytes on other plants or objects. Bromeliads get much of their water and nutrients from rainfall and debris trapped between their tightly packed leaves. Other examples of epiphytes include orchids and staghorn ferns.

FIGURE 19 The cup-shaped structure on this liverwort is a gemmae cup. Gemmae cups are adaptations for asexual reproduction. The coin-shaped gemmae in the bottom of the cup are splashed out of the cup by raindrops. Each gemma that lands in a suitable environment forms a new gametophyte generation of the plant.

PLATE IX

FIGURE 20 Puffballs are fungi that belong to the division Basidiomycota. The puffballs in this photo are releasing millions of tiny spores, each of which can form a new organism.

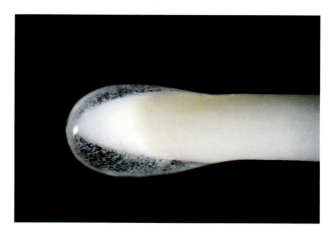

FIGURE 21 Tips of roots secrete large amounts of mucilage, a lubricant that helps the root force its way through the soil. The mucilage is secreted primarily by the root cap. Movement of the root through soil is also aided by the sloughing of root cap cells. These sloughed cells are visible in the drop of mucilage on the tip of this root.

FIGURE 22 This flower beetle is gathering pollen from an oxeye daisy. In the process, the beetle's body becomes covered with pollen. This pollen is transferred to other flowers when the beetle visits other flowers, thereby promoting cross-pollination.

FIGURE 23 Carnivorous plants such as this Venus's-flytrap live in nitrogen-poor soil. The ability of these plants to trap and digest insects is an adaptation to gather nitrogen. Leaves of Venus's-flytrap are modified into hinged traps that catch small insects. While this fly crawls across the surface of the trap searching for nectar, it will probably touch the trigger hairs on the surface of the trap. When the fly has touched two of these hairs (or one hair twice), the trap will grow shut, trapping the fly. The fly is then digested by enzymes secreted by the trap.

PLATE X

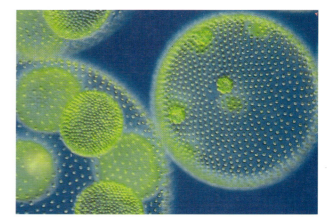

FIGURE 25 *Volvox* is a colonial green alga that grows in fresh water. These organisms are made up of thousands of cells organized into a beachball-like sphere. Some cells of *Volvox* are specialized for reproduction. The aggregates of cells on *Volvox* are new organisms developing from these reproductively specialized cells.

FIGURE 24 The droplets of water at the edges of these leaves of a strawberry plant are formed by guttation. Guttation results from root pressure (i.e., a positive pressure in the xylem) common in small plants growing in moist soil on cool, damp mornings. Although most water movement in plants occurs as a result of evaporation of water from leaves, guttation may be an important way of moving water in short, herbaceous plants.

FIGURE 26 Bladderwort is an aquatic plant that traps and digests protozoa and mosquito larvae. Some of the submerged leaves of bladderwort are modified into hollow traps. Organisms that touch the trigger hairs at the trap doors are sucked into the trap, where they are digested.

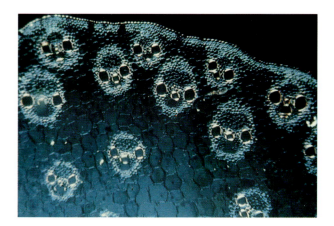

FIGURE 27 The monkeyface-shaped structures in this cross section of a corn stem are vascular bundles. These bundles consist primarily of xylem, a tissue specialized for long-distance transport of water, and phloem, a tissue specialized for long-distance transport of organic solutes. The evolution of these vascular tissues increased the efficiency of movement of water and solutes from the soil to distant leaves, thereby enabling land plants to grow taller and absorb more light.

PLATE XI

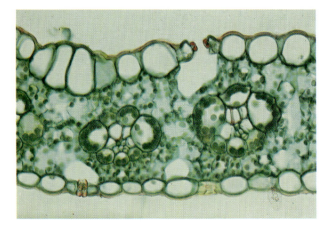

FIGURE 28   The large, chloroplast-containing cells surrounding the vascular tissue in this cross section of a corn leaf are bundle sheath cells. In most plants, bundle sheath cells are not photosynthetic. However, in corn and other C4 plants, the bundle sheath cells contain chloroplasts and are photosynthetic. This unique leaf anatomy is associated with C4 photosynthesis, an adaptation of many tropical plants to hot, dry conditions. Carbon dioxide used in photosynthesis enters the leaf through stomata such as the one shown on the upper surface of this leaf.

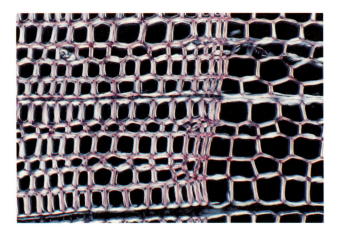

FIGURE 29   The wood of gymnosperms (such as this pine) consists almost exclusively of tracheids. These water-conducting cells are relatively small and help to support the plant. Although water moves slower through tracheids than through vessel elements, tracheids are less likely to be disabled by air bubbles that form in response to freezing and wind-induced bending of branches.

The larger cells at the left of this photo form during the wet days of spring and are called spring wood; the smaller cells at the right form during the drier days of summer and are called summer wood. The change in density between spring and summer wood produces a growth ring, which appears as grain in wood.

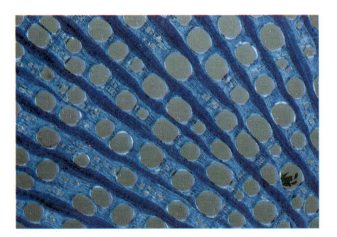

FIGURE 30   Unlike the xylem of gymnosperms, which contains only tracheids, the xylem of angiosperms also contains vessel elements. These vessel elements are much wider than tracheids and appear in this picture as large circles. Vessels are an adaptation for increased rates of water flow in angiosperms.

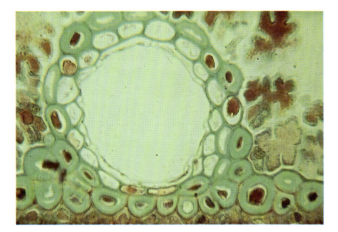

FIGURE 31   This resin duct from a leaf of a pine contains resin. Resin ducts are also found in the wood and bark of pine. When the plant is damaged, resin oozes out of the duct, thereby sealing the wound, minimizing infection and deterring herbivores.

PLATE   XII

FIGURE 32 The biggest problem facing plants growing in dry habitats such as deserts is desiccation. Window plants such as *Frithia* cope with the harsh desert by forming succulent leaves with transparent tips. Only the tips of the leaves protrude from the soil, thereby minimizing the amount of surface area exposed to herbivores and the desert's dry air. Light reaches the underground photosynthetic tissue through the transparent, window-like tip of the plant's leaves.

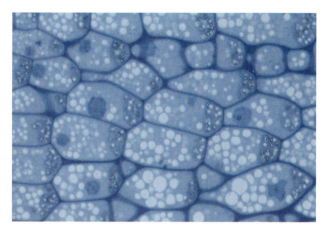

FIGURE 33 Roots perceive and respond to a variety of environmental stimuli, including light, moisture, and gravity. The root cap is the organ that senses these stimuli. This longitudinal section of a root cap of corn shows columella cells (named for their columnar arrangement). Each columella cell contains large amyloplasts that have settled to the bottom of the cell. The movement of these amyloplasts to the bottom of columella cells is thought to be how roots perceive gravity.

FIGURE 34 The showy flowers on these cacti are adaptations for reproduction. The flowers attract animals that transfer pollen from flower to flower, thereby promoting cross pollination.

FIGURE 35 Temperate plants must sense and prepare for changes in season if they are to survive. These winter buds of buckeye are adaptations for surviving the cold winter. The delicate meristems are covered by tough, thick, bud scales that protect the meristem from winter's cold.

PLATE XIII

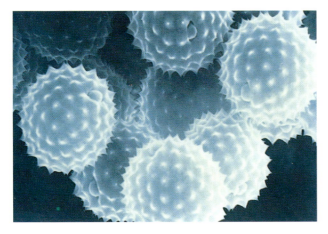

FIGURE 36 Pollen grains of ragweed are spherical and elaborately sculptured. Their outer wall contains many proteins that regulate germination of the pollen grain when it lands on a stigma of a ragweed flower. These same proteins also cause allergies in many people.

FIGURE 37 The yellow and red colored trees in this deciduous forest will soon shed their leaves in preparation for winter. The leaves of these trees have lost their usual green color because of the breakdown of chlorophyll. This breakdown and recycling of chlorophyll unmasks other pigments, accounting for the spectacular changes in colors that occur in autumn. The shedding of leaves before winter reduces the surface area of the plant, thereby reducing desiccation.

FIGURE 39 Mistletoe is a parasitic plant that infects a variety of trees. Although mistletoe is green, it is not photosynthetic. The parasitic lifestyle of mistletoe and other plants is an adaptation for securing water and nutrients from the host plant.

FIGURE 38 The shoot apical meristem produces the stems and leaves that form the shoot of a plant. Unlike the root apical meristem, which is protected by a root cap, the shoot apical meristem is protected by young leaves that arch over the meristem. Axillary buds form at the base of each new leaf. These buds can assume the function of the shoot apical meristem if the apical meristem is damaged. Each bud can also form a branch or a flower.

PLATE XIV

FIGURE 40 These are flowers of Indian pipe, a parasitic plant. Unlike most other parasitic plants, Indian pipe parasitizes the mycorrhizae associated with nearby trees. The only time Indian pipe appears above ground is when it flowers.

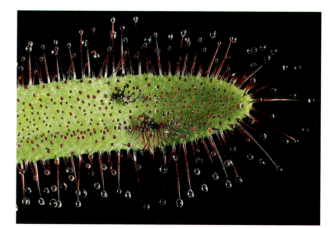

FIGURE 41 Sundew is a carnivorous plant that traps small insects. Its club-shaped leaves have many glandular hairs that secrete a sticky nectar. Insects such as those on this leaf become trapped in the nectar. In some species of sundew, the leaf then folds into a trap surrounding the insect, after which the insect is digested by enzymes secreted by the leaf. The leaf absorbs and uses nutrients released from the insect's body.

FIGURE 42 Leaves of pitcher plants are modified into tubular vats. These vats contain a mixture of digestive enzymes and nectar that attracts insects. Insects crawling along the downward-pointing hairs lining the inside of the pitcher fall into the vat of enzymes, where they are digested. The plant then absorbs and uses nutrients released from the insect's bodies. Interestingly, some pitcher plants contain more than a liter of digestive fluid.

FIGURE 43 The majestic Saguaro cactus is common in Arizona's deserts. These succulent plants conserve water by reducing their leaves to spines and by producing thick stems that store large amounts of water.

PLATE XV

FIGURE 44 Spanish moss, a plant common throughout the southeastern United States, is a flowering plant, not a moss. The Spanish moss shown here living as an epiphyte on cypress trees lacks roots; it absorbs water and nutrients through hairs on its stems and leaves.

FIGURE 45 Mutations produce new alleles and genetic combinations that may be adaptive. One branch of this peppermint peach tree has grown from a meristem that had a simple mutation for red petals. If this mutation is adaptive it will probably persist in the gene pool of subsequent generations.

FIGURE 46 Angiosperms disperse their seeds by encasing them in fruit. These fruits, which often contain nutrients such as sugar and starch, are often eaten by animals. The seeds of the fruit move throughout the animal's digestive tract and are deposited in the animal's feces in a new location, thereby enabling the plant to invade a new territory.

FIGURE 47 Leaves of plants such as this ocotillo are reduced to spines, thereby minimizing the amount of surface area exposed to the environment. The spines also help protect the plant from hungry animals.

PLATE XVI

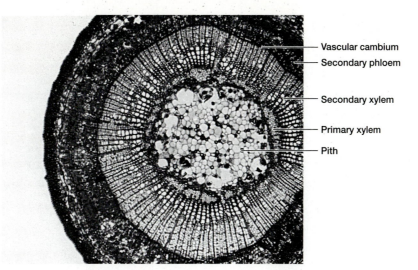

Vascular cambium
Secondary phloem
Secondary xylem
Primary xylem
Pith

FIGURE 17.6

One-year-old basswood stem.

3. Examine a prepared slide of secondary xylem of pine (*Pinus*). In gymnosperms such as pine, the conducting cells of xylem are all **tracheids.** The absence of vessel elements gives wood of these plants a relatively uniform appearance.

4. Now examine a prepared slide of secondary xylem of oak (*Quercus*), an angiosperm (i.e., flowering plant).

Question 6

a. How does wood of pine differ from that of oak?

_____

b. What are the large cells in oak wood? _____

What is their function? _____

_____

c. Which type of wood do you think transports more

water per unit area, wood of pine or oak? _____

Why? _____

_____

Bark

Bark includes all tissues outside of the vascular cambium, including the secondary phloem (fig. 17.7). Secondary phloem consists of pyramidal masses of thick- and thin-walled cells. The thin-walled cells are the conducting cells.

The increase in stem circumference resulting from activity of the vascular cambium eventually ruptures the epidermis. The ruptured epidermis is replaced by a tissue called the periderm that, like the epidermis, functions to minimize water loss. **Periderm** includes cork cells produced by another secondary meristem called the **cork cambium.**

Procedure 17.6
Examine bark

1. Locate the cork cambium in the cross section of 1-, 2-, and 3-year-old basswood stems. The cork cambium is a band of thin-walled cells located beneath the epidermis. The cork cambium produces cork cells to the outside and cork parenchyma to the inside. Cork cells stain red because of the presence of suberin, a water-impermeable lipid.

Question 7

a. Is the amount of cork similar in 1-, 2-, and 3-year-old

stems? _____

What do you conclude from your observation?

_____

2. As the stem diameter continues to increase, the original periderm ruptures and new periderms form in the underlying tissues. Tissues outside the new periderm die and form encrusting layers of bark. Gas exchange through peridermal tissues occurs through structures called **lenticels** (fig. 17.8). Examine a prepared slide of a lenticel.

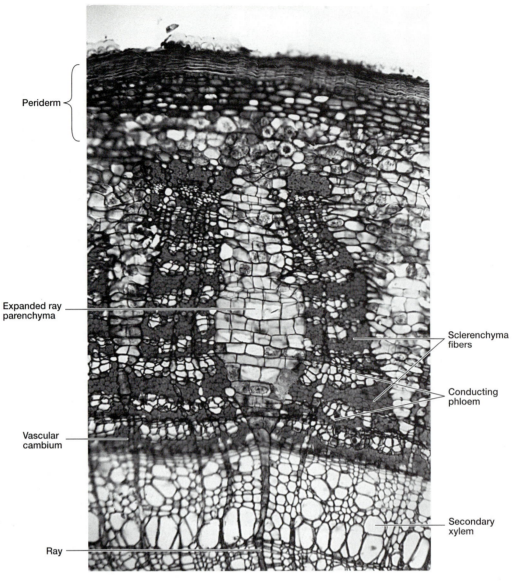

Periderm

Expanded ray
parenchyma

Sclerenchyma
fibers

Conducting
phloem

Vascular
cambium

Secondary
xylem

Ray

FIGURE 17.7

Bark, including secondary phloem and fibers.

Question 8

a. How does a lenticel differ from the remainder of the

periderm? _____

_____

How does this difference relate to the function of

lenticels? _____

_____

Summary of Activities

1. Examine the shoot tip of *Coleus*.
2. Examine the external features of a woody stem.
3. Study the primary tissues of stems of monocots and dicots.
4. Examine the secondary tissues of woody stems, including bark and lenticels.

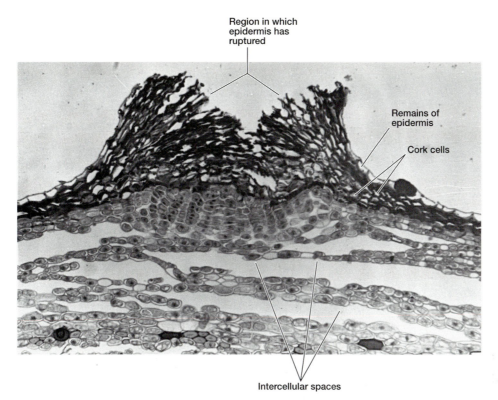

Region in which
epidermis has
ruptured

Remains of
epidermis

Cork cells

Intercellular spaces

FIGURE 17.8

Lenticel.

Questions for Further Thought and Study

1. Why does a stem typically contain more sclerenchyma and collenchyma than a leaf?
2. How is the internal anatomy of a stem different from that of a root?
3. What is the significance (i.e., adaptive value) of secondary growth?

Writing to Learn Botany

Would you expect to find annual rings in wood of a tropical dicot tree? Why or why not?

Leaves

18

Objectives

By the end of this exercise you will be able to

1. describe the different types of venation of leaves
2. describe the internal anatomy of leaves of dicots and monocots
3. understand the significance of anatomical differences in leaf anatomy
4. describe the adaptations of leaves of mesophytes, xerophytes, and hydrophytes
5. understand the structural basis for leaf abscission.

With few exceptions, photosynthesis in most plants occurs in leaves. Leaves typically consist of a **blade** and a **petiole.** The petiole attaches the leaf blade to the stem (see plate VI, figs. 10 and 11). **Simple leaves** have one blade connected to the petiole, while **compound leaves** have several **leaflets** sharing one petiole (fig. 18.1). **Palmate** leaflets of a compound leaf arise from a central area, as your fingers arise from your palm. **Pinnate** leaflets arise in rows along a central midline. Examine the simple and compound leaves on display, and sketch at least one of each type of leaf on a separate sheet of paper.

Leaves are also classified according to their **venation,** or arrangement of veins (fig. 18.2). **Parallel veins** extend the entire length of the leaf with little or no cross-linking. **Pinnately veined** leaves have one major vein (midrib) from which other veins branch. **Palmately veined** leaves

have several main veins each having branches. Veins of vascular tissue in leaves are continuous with vascular bundles in stems.

The arrangement of leaves on a stem is called **phyllotaxis** and characterizes individual plant species (fig. 18.3). **Opposite phyllotaxis** refers to two leaves per node located on opposite sides of the stem. **Alternate phyllotaxis** refers to one leaf per node, with leaves appearing first on one side of the stem and then on another. **Whorled phyllotaxis** refers to more than two leaves per node. Examine the plants on display in the lab, and determine their phyllotaxis.

Procedure 18.1
Observe venation of leaves

Examine the leaves on display in the lab, and determine their venation. Be sure to examine at least one dicot (e.g., *Coleus*) and one monocot (e.g., *Zebrina*). Sketch a few of these leaves to show their venation. List the names of common leaves on demonstration and indicate whether each is simple, pinnately compound, or palmately compound. Also indicate whether venation in each leaf is parallel, pinnate, or palmate.

Question 1
a.   How does venation differ in these plants? _____
_____

Internal Anatomy of Leaves

Botanists use several criteria to classify foliage leaves. For example, foliage leaves may be classified as simple or compound based on their morphology (see fig. 18.1). Foliage leaves also can be classified based on their anatomical adaptations to the environment. Thus, leaves of meso-

phytic plants are adapted to a "normal" water supply, leaves of xerophytic plants are adapted to dry environments, and leaves of hydrophytic plants are adapted to aquatic or semiaquatic environments. Although this may seem like a rather artificial classification scheme to use in a botany course, many characteristics contribute to this type of classification.

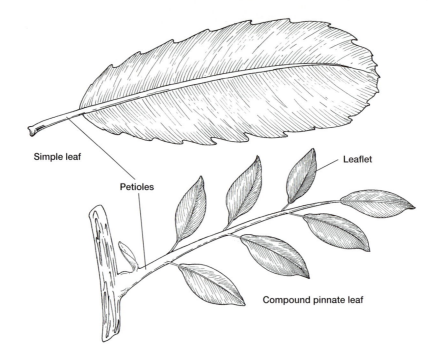

FIGURE 18.1

Simple and compound leaves.

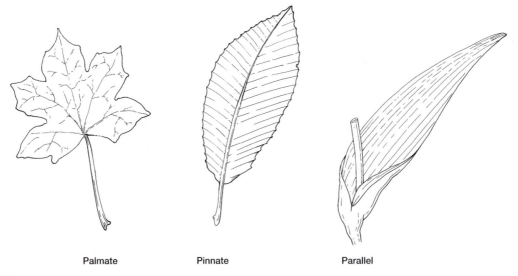

Palmate   Pinnate   Parallel

FIGURE 18.2

Venation of leaves.

Procedure 18.2
Examine leaves of a mesophyte

Dicot

1. Examine a cross section of a leaf of *Ligustrum*, a dicot known as privet (fig. 18.4).

2. Note that the leaf is only ten to fifteen cells thick. Locate pores in the epidermis called **stomata,** each of which is surrounded by two guard cells (you will study stomata again in exercise 22). Just below the upper epidermis are closely packed cells called **palisade mesophyll** cells. These cells contain about fifty chloroplasts per cell. Below the palisade layers are **spongy mesophyll** cells with numerous intercellular spaces.

18-2

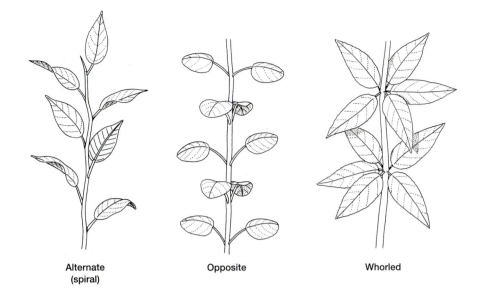

Alternate (spiral)   **Opposite**   **Whorled**

FIGURE 18.3

Patterns of leaf arrangement.

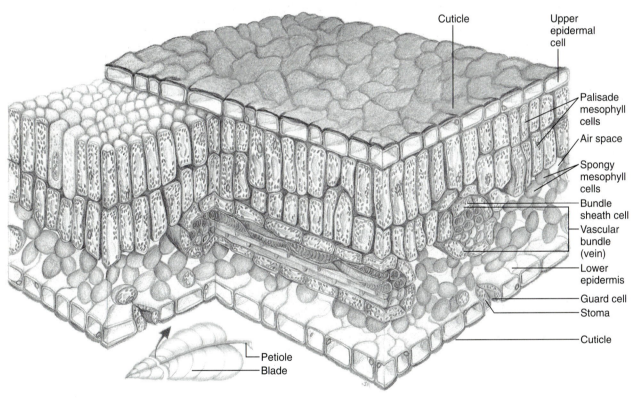

Cuticle

Upper epidermal cell

Palisade mesophyll cells

Air space

Spongy mesophyll cells

Bundle sheath cell

Vascular bundle (vein)

Lower epidermis

Guard cell

Stoma

Cuticle

Petiole

Blade

FIGURE 18.4

Ligustrum leaf, cross section. Most photosynthesis occurs in the densely packed palisade mesophyll cells, which are just beneath the upper epidermis of the leaf. Gas exchange occurs through stomata, which are usually most abundant on the lower side of the leaf. Water loss is minimized by the waxy cuticle that covers the leaf.

Question 2
a. What is the function of stomata? _____

_____

b. Do epidermal cells of leaves have a cuticle? (Cuticle is a waterproof layer secreted on the outer surface of the epidermal cells.) _____

_____

c. What is the significance of chloroplasts being concentrated near the upper surface of the leaf?

_____

d. What is the significance of the air spaces near the lower surface of the leaf? _____

_____

e. What tissues form a vein? _____

_____

Monocot
3. Examine and sketch a cross section of a leaf of corn (*Zea mays*).

Question 3
a. How does the internal anatomy of a corn leaf differ from that of a leaf of *Ligustrum*? _____

_____

b. What is the functional significance of this difference?

_____

_____

c. Based on the arrangement of vascular tissue, how could you distinguish the upper and lower surfaces of the leaf? _____

_____

Gymnosperm
Pine (*Pinus*) is a gymnosperm whose leaves are shaped like needles.
4. Examine a cross section of a pine needle.
5. Identify the epidermis, mesophyll, and vascular tissue.

Question 4
a. What are the functions of each of these tissues?

_____

b. What features of a pine leaf suggest that *Pinus* is a xerophytic plant? _____

_____

c. How would you describe the mesophyll of this leaf?

_____

XEROPHYTIC LEAVES

Many characteristics are regarded as xerophytic, and an endless number of leaf variations occur in dry habitats. We will consider only two examples: (1) a leaf reduced in size to minimize water loss, and (2) a leaf designed to store water.

Procedure 18.3
Examine leaves of a xerophyte
1. Obtain a prepared slide of a cross section of a leaf of *Ammophila arenaria* (beach grass).

Question 5
a. How is the leaf shape related to water conservation?

_____

b. Compare the upper and lower surfaces of the leaf. How do they differ? _____

_____

Of what significance is this? _____

_____

2. Observe the cells at the tips of ridges between the surface grooves.

Question 6
a. Can you imagine that these ridges might fit together under certain conditions? _____

How would this help the plant conserve water? ____

_____

b. What is the structure of the mesophyll? _____

_____

3. Obtain a prepared slide of a succulent leaf of *Aloe*.

Question 7
a. Where are the parenchyma cells located? _____

_____

b. What type of tissue constitutes the remainder of the mesophyll? _____

c. How many veins are present? _____
d. Are they small, large, or unusual in any respect?

_____

4. Study the *Aloe* epidermis.

Question 8
a. Where are the stomata located? _____
b. What is the nature of the cuticle? _____

_____

c. Why do we regard this leaf as xerophytic? _____

_____

Procedure 18.4
Examine leaves of a hydrophyte

1. To study a floating leaf, obtain a slide of *Nymphaea* (water lily) leaf, x.s. Notice the modifications of the palisade tissue and the substomatal cavities.

Question 9
a. On which epidermis can you locate stomata?

_____

Of what significance is this? _____

_____

b. How much xylem versus phloem occurs in the

vascular bundles? _____
Of what significance is this?

_____

2. Obtain a slide of *Typha latifolia* (cattail) leaf. Cattail is considered semiaquatic because it grows along pond edges or in marshy areas. The leaves are several feet long, oriented vertically, and resemble swords.

Question 10
a. What special modifications exist in this leaf? _____

_____

b. Do the long leaves have any supportive tissue? _____

_____

c. How could you have deduced that these leaves are

oriented vertically? _____

_____

LEAF ABSCISSION

Abscission layers occur in various locations in different plants, including the base and top of the leaf petiole, at the base of small branches, at the base of many flower parts, and in the stalks of fruits. Because the anatomy of most abscission layers is similar, you will study only one example, an abscission layer at the base of a leaf petiole.

Procedure 18.5
Observe leaf abscission

1. Examine a prepared slide of *Prunus* leaf abscission, l.s.
2. Identify these tissues: abscission layer, separation layer, and suberized tissue. In older, senescing leaves,

the developing abscission zone may be visible to the naked eye as a lighter-colored band near the base of the petiole.

Question 11
a. What is the function of each tissue? _____

_____

b. What happens to the vascular tissue during abscission?

_____

Summary of Activities

1. Examine the types of venation in leaves of monocots and dicots.
2. Examine the structure of simple and compound leaves.
3. Examine the internal structure of leaves of monocots, dicots, and gymnosperms.
4. Examine the structural adaptations of leaves of xerophytic and hydrophytic plants.
5. Examine the structure of the abscission zone of leaves.

Questions for Further Thought and Study

1. What tissues form a vein? What is the function of each of these tissues?
2. What is the significance of the structural differences between the spongy and palisade mesophyll layers of leaves?
3. What are some of the adaptations characteristic of xerophytic leaves? What is the functional significance of each?
4. What are some of the adaptations characteristic of hydrophytic leaves? What is the functional significance of each?
5. Do leaves of all plants have an abscission layer? If not, what happens to the leaves on such plants?

Doing Botany Yourself

Choose a couple of defined environments nearby, such as a vacant field or riverside. Survey the variety of leaf morphologies of the dominant plants. What can you conclude about the predominance of monocots versus dicots in each environment?

Writing to Learn Botany

Why is leaf abscission especially important for temperate plants?

Embryo and Seedling Development

Objectives

By the end of this exercise you will be able to

1. describe the major stages of embryo development
2. understand the mechanism governing the tetrazolium test for seed viability
3. describe epigean and hypogean development in seedlings of bean, pea, and corn.

EMBRYO DEVELOPMENT

A seed is a mature ovule and includes a seed coat, a food supply, and an embryo. Development of the mature zygote occurs within the seed and before germination. This embryology and its control are complex, but various stages of development are easily observed. The stages of embryo development in the seed of *Capsella* are illustrated in figure 19.1. The developing embryo grows, absorbs endosperm, and stores nutrients in "seed leaves" called cotyledons. Development includes the following stages:

*   *Proembryo.* During development, the **zygote** divides to form a mass of cells called the **embryo.** Initially the embryo consists of a **basal cell,** a **suspensor** and a two-celled **proembryo.** The suspensor is the column of cells that pushes the embryo into the **endosperm.** The endosperm is extensive but is digested during development of the embryo.

*   *Globular Stage.* Cell division of the proembryo soon produces the globular stage, which is radially symmetrical and has relatively little internal cellular organization.

*   *Heart-shaped Stage.* Differential division of the globular stage produces bilateral symmetry and two **cotyledons** that form a heart-shaped embryo. The enlarging cotyledons store digested food from the endosperm. Tissue differentiation begins, and root and shoot meristems form.

*   *Torpedo Stage.* The cotyledons and root axis elongate to produce an elongate torpedo stage embryo. Procambial tissue appears; this tissue later develops into vascular tissue.

*   *Mature Embryo.* The mature embryo has large, bent cotyledons on each side of the **stem apical meristem.** The **radicle,** later to form the root, differentiates toward the suspensor and includes a **root apical meristem** and **root cap.** The **hypocotyl** is the region between the apical meristem and the radicle. The endosperm is now depleted, and food is stored in the cotyledons. The **epicotyl** is the region between attachment of the cotyledons and the stem apical meristem. It does not elongate in the mature embryo stage.

Procedure 19.1
Examine development of a *Capsella* embryo

1. Obtain and examine prepared slides showing the various stages of embryo development in *Capsella*.

2. Locate examples of globular, heart-shaped, torpedo, and mature embryos. These stages are continuous,

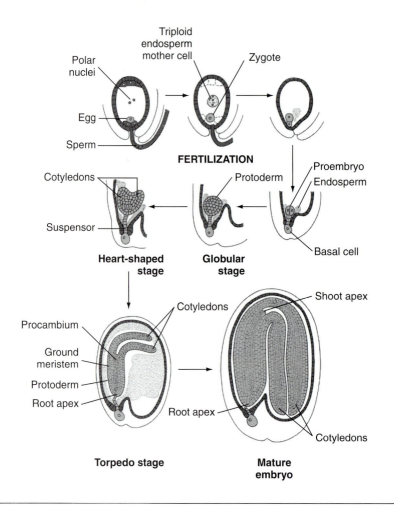

FIGURE 19.1

Stages of development in a *Capsella* embryo. This development transforms a zygote into a mature embryo.

so each slide may have multiple stages and intermediate phases between stages.

3. Compare your observations with those shown in figure 19.1.

Question 1

a. Why is the endosperm being digested? _____

_____

b. Is *Capsella* a monocot or a dicot? _____

How can you tell? _____

_____

Procedure 19.2
Examine a corn grain with embryo

1. Examine a prepared slide of a corn grain (fig. 19.2). Identify the following features:

endosperm

cotyledon (scutellum)

coleoptile (sheath enclosing shoot apical meristem and leaf primordia of grass embryos)

root

root cap

coleorhizae (sheath enclosing embryonic root of grass embryo)

shoot apical meristem

2. Use a razor blade to longitudinally split a water-soaked corn grain.

3. Add a drop of iodine to the cut surface and observe the staining pattern. Indicate this pattern by shading the appropriate area on the photograph of a corn grain (fig. 19.2).

Question 2

a. What does this staining pattern tell you about the

content of the endosperm and embryo? _____

_____

b. Do mature seeds of monocots or dicots store most of

their food in cotyledons? _____

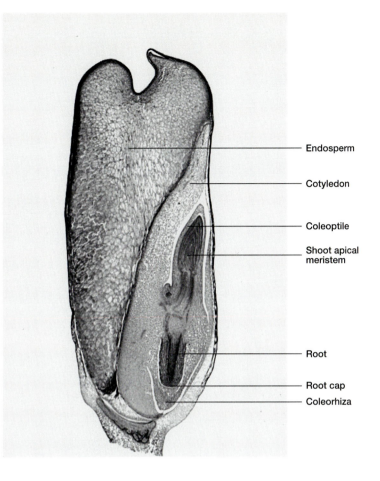

FIGURE 19.2

The structure of a corn (*Zea mays*) grain, longitudinal section. The embryo is nourished by the starchy endosperm.

SEED VIABILITY

Seed **viability,** or ability to germinate, varies and can be assessed with a **tetrazolium test.** During germination, aerobic respiration increases dramatically and electrons are transferred by various dehydrogenase enzymes. Dehydrogenases can also transfer electrons to compounds not found in the seed. One of these compounds, 2,3,5-triphenyltetrazolium chloride, can be reduced by dehydrogenases and turns from colorless to red. Reduced tetrazolium indicates active respiration (and therefore viability) of cells. More red color of a seed exposed to tetrazolium indicates greater respiration. Dead cells do not reduce tetrazolium. In this manner, a tetrazolium test of a sample of seeds will measure the viability of that group or lot of seeds.

Procedure 19.3
Test seed viability using the tetrazolium test

1.   Soak forty corn seeds in water overnight.
2.   Boil twenty of the seeds in water for 20 minutes.

3.   Section five of the living and five of the boiled seeds. To do this, place each seed flat with the scutellum up and cut medially with a razor blade. This should section through the embryo and endosperm.
4.   Place the seed sections in a petri plate with 1% tetrazolium chloride.
5.   Observe any color change in the seeds every 20 minutes for 1 hour.
6.   Record your observations in table 19.1.

Question 3
a.   Did the boiled seeds show signs of respiration? _____

Why or why not? _____

b.   Is the tetrazolium test the definitive test for seed

viability? _____

Why or why not? _____

c.   Does the presence of respiration mean that the seed

will germinate? _____

Why or why not? _____

TABLE 19.1

RESULTS OF TETRAZOLIUM TEST FOR SEED VIABILITY

Seed Type	Observations		
	20 minutes	40 minutes	60 minutes
Boiled Seed			
Living Seed			

SEEDLING DEVELOPMENT

Seedling development begins when the seed coat ruptures. The growing embryo breaks the seed coat, and the young plant emerges. The endosperm and cotyledons supply food for the young seedling until it establishes a root system and photosynthetic tissue.

During **epigean development** the cotyledons are pushed above the soil surface by the growing hypocotyl. The hypocotyl is the region between the radicle (embryonic root) and the attachment of the cotyledons. In **hypogean development** the cotyledons remain below the soil surface.

Procedure 19.4
Examine seedling development in bean, pea, and corn

1. Obtain bean, pea, and corn seedlings at various stages of development. The seedlings should be marked by hours or days of development.

2. Examine the developing bean seedlings; compare them with the diagrams shown in figure 19.3. Identify the following structures:

   seed coat   hypocotyl
   secondary root   leaves
   radicle   cotyledons
   root apical meristem   epicotyl
   stem apical meristem   primary root

3. Examine the developing pea seedlings and locate the following structures on the seedlings:

   seed coat   primary root
   cotyledons   secondary root
   radicle   epicotyl
   root apical   leaves
   stem apical meristem

   In the space below, draw a pea seedling and label it appropriately.

4. Examine the developing corn seedlings; compare them with the diagrams shown in figure 19.4. Locate the structures listed. The epicotyl of a corn seedling is surrounded by a sheath called a **coleoptile.**

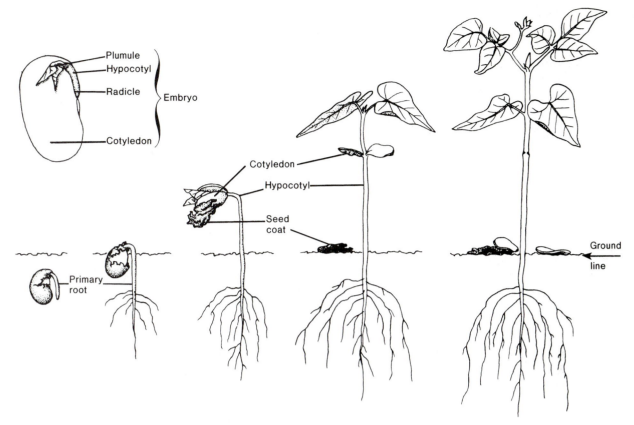

FIGURE 19.3

Bean seed and seedling development.

Adventitious roots develop from the stem, and the primary root degenerates. **Prop roots** may develop from the stem aboveground.

seed coat   primary root
cotyledon   secondary root
endosperm   epicotyl
radicle   leaves
stem apical meristem   root apical meristem
prop root   coleoptile

Question 4

a. Which of the three plants—bean, pea, or corn—has epigean development? _____

b. Which has hypogean development? _____
_____

c. Are cotyledons photosynthetic? _____
In which plants? _____
_____

d. Which part of the embryo emerges first in the bean seedling? _____

pea seedling? _____
corn seedling? _____
What is the importance of this part emerging first?
_____

e. What ultimately happens to the cotyledons of the bean? _____
_____

f. How can a corn seedling survive with such a small cotyledon? _____

g. How long did it take the epicotyl to emerge in each type of seedling? _____

h. How long would you estimate before the seedling is nutritionally independent of the endosperm or cotyledon in each type of seedling? _____
_____

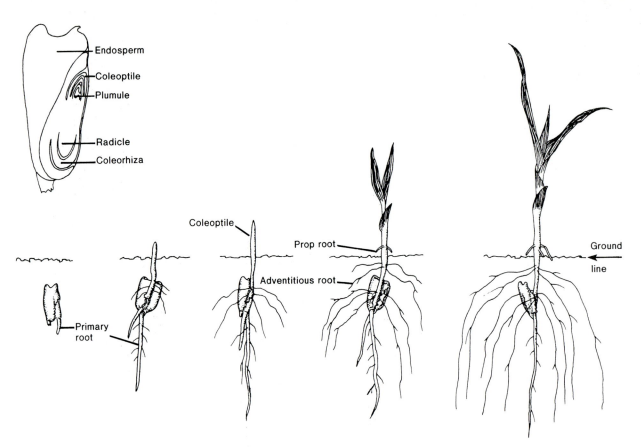

FIGURE 19.4

Corn seed and seedling development.

Summary of Activities

1. Examine prepared slides showing the major stages of *Capsella* embryo development.
2. Examine parts of a corn embryo.
3. Perform the tetrazolium test for seed viability.
4. Examine bean, pea, and corn seedlings with epigean and hypogean development.

Questions for Further Thought and Study

1. Interestingly, some seeds must pass through the digestive tract of an animal before they can germinate. What is the adaptive significance of this?
2. What events mark the establishment of a seedling from a germinating seed?

Doing Botany Yourself

Germinate squash, peanut, and sunflower seeds. Examine seedling development and determine if it is epigean or hypogean.

Writing to Learn Botany

Describe the advantages and relative disadvantages of large versus small seeds.

Plant Growth Regulators

Objectives

By the end of this exercise you will be able to

1. explain the modes of action of auxin and gibberellic acid

2. explain the concept of a bioassay

3. describe how to use a bioassay to approximate the concentration of a physiologically active substance.

Plant growth and development are controlled by internal chemical signals called **growth regulators.** In this exercise you will study some common physiological responses of plants to growth regulators such as auxin and gibberellic acid (fig. 20.1). Remember that each of these responses is part of an overall design for survival and reproduction in varied environments.

Botanists examine the effects of growth regulators on plants and employ techniques to measure the concentrations of these substances. These techniques are especially important because most regulators are active in extremely low concentrations and are difficult to measure within plant tissues. To relate chemical concentration to real biological processes, botanists often use a technique called **bioassay.**

Bioassay is a method to determine the biological activity of a physiologically active substance. This method directly measures an organism's response to a chemical, which can be of any chemical nature, such as a nutrient, herbicide, or pesticide. The biological activity elicited by the chemical being bioassayed is then compared to the activity elicited by a series of known concentrations of the chemical. Then we can deduce the chemical's unknown concentration.

PLANT GROWTH REGULATORS

Indoleacetic Acid (IAA)

Indoleacetic acid (IAA), also known as **auxin,** is the best-known growth regulator in plants. It is produced by shoot tips and dramatically affects cell growth (fig. 20.2). Auxin also inhibits development of axillary buds. That is, IAA promotes apical dominance.

Procedure 20.1
Examine the effects of auxin

1. Examine the *Coleus* plants subjected to the treatments described in table 20.1.

2. Record your observations in table 20.1.

Question 1

a. What is the effect of replacing the shoot tip with IAA? _____

b. Why was it necessary to apply only lanolin to one plant? _____

2,4-Dichlorophenoxyacetic Acid

2,4-dichlorophenoxyacetic acid (2,4-D) is a synthetic auxin used frequently in agriculture as a herbicide. When applied at high concentrations, 2,4-D disrupts the physiology of many plants.

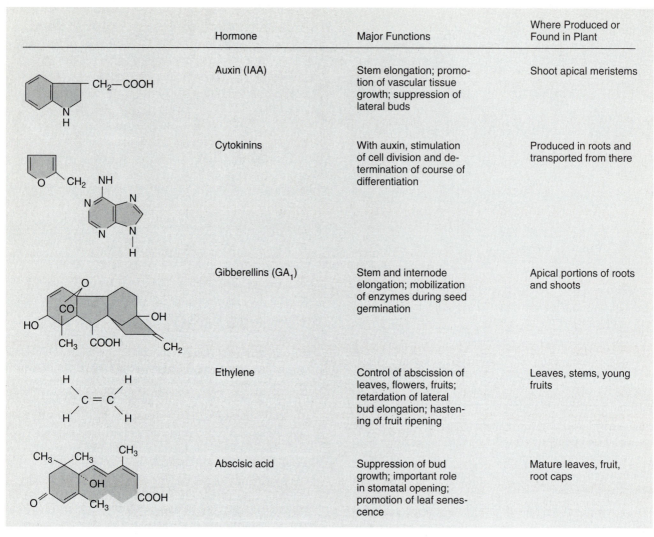

	Hormone	Major Functions	Where Produced or Found in Plant
	Auxin (IAA)	Stem elongation; promotion of vascular tissue growth; suppression of lateral buds	Shoot apical meristems
	Cytokinins	With auxin, stimulation of cell division and determination of course of differentiation	Produced in roots and transported from there
	Gibberellins (GA$_1$)	Stem and internode elongation; mobilization of enzymes during seed germination	Apical portions of roots and shoots
	Ethylene	Control of abscission of leaves, flowers, fruits; retardation of lateral bud elongation; hastening of fruit ripening	Leaves, stems, young fruits
	Abscisic acid	Suppression of bud growth; important role in stomatal opening; promotion of leaf senescence	Mature leaves, fruit, root caps

FIGURE 20.1

Functions of the major plant growth regulators. Plant growth regulators affect a variety of responses in plants, including abscission, cellular elongation, and bud growth.

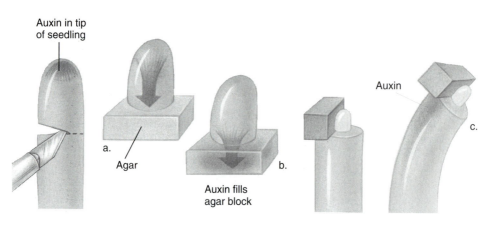

FIGURE 20.2

Demonstrating the action of auxin. (*a*) The tip of a grass seedling is removed and placed on an agar block. Auxin produced in the tip diffuses into the agar. (*b*) The block of agar is put on one side of the cut end of another grass seedling. (*c*) The seedling bends away from the side with the agar block because auxin promotes elongation of cells in shoots of grass seedlings.

20-2

TABLE 20.1

EFFECTS OF AUXIN ON COLEUS PLANTS

Treatment	Observations
Control (no treatment)	
Shoot tip removed	
Shoot tip removed; cut surface coated with 1% 1AA in lanolin	
Shoot tip removed; cut surface coated with lanolin alone	

Procedure 20.2
Examine the effects of 2,4-D

1. Examine plants that were treated 2 weeks ago with 2,4-D. The pots originally contained a mixture of monocots (e.g., corn, oats) and dicots (e.g., peas, beans).
2. Record your observations in table 20.2.

Question 2

Which plants were affected most by the herbicide?

_____

Why is this important? _____

_____

Gibberellic Acid (GA)

Dwarfism often results from a plant's inability to make active forms of **gibberellic acid (GA),** another plant growth regulator. Gibberellic acid promotes stem elongation and mobilizes enzymes during seed germination in some species (fig. 20.3).

Procedure 20.3
Examine the effects of gibberellic acid

1. Examine the four trays of corn plants that were treated in the following ways:

    Tray 1: Normal plants, untreated

    Tray 2: Dwarf plants, untreated

    Tray 3: Normal plants, treated with 2 to 3 drops of GA on alternate days for 2 weeks

    Tray 4: Dwarf plants, treated with 2 to 3 drops of GA on alternate days for 2 weeks

2. Record your observations in table 20.3.

Question 3

a. What is the effect of GA on genetic dwarfism?

_____

_____

b. Did applying GA to normal plants have any effect?

_____

If so, what? _____

_____

TABLE 20.2

EFFECTS OF 2,4-D ON MONOCOTS AND DICOTS

Type of Plant Treated with 2,4-D	Observations
Monocots	
Dicots	

FIGURE 20.3

Effect of gibberellic acid on cell elongation in cabbage. Cabbage (*Brassica oleracea*), a biennial that is native to the seacoasts of Europe, will "bolt" when the heads are treated with gibberellin. Although more than 80 gibberellins have been isolated from natural sources, apparently only one is active in shoot elongation.

BIOASSAY FOR PLANT GROWTH REGULATORS

In the following procedures you will bioassay the level of activity of three plant growth regulators: auxin, 2,4-D, and gibberellic acid. You will assay the effect of auxin on curvature of split stems, the effect of 2,4-D on root elongation, and the effect of gibberellic acid on enlargement of radish cotyledons. All of the bioassays will be done at a constant pH. This stabilizes the reactivities and potency of the regulators.

Auxin

You will measure how much a stem curves in response to known concentrations of auxin. Using this data you will construct a standard curve (graph) showing the relationship between known concentrations of auxin and degree of curvature of a stem. You will then bioassay a solution of auxin of unknown concentration by measuring its effect on stem curvature. You can then use the standard curve to determine the unknown concentration of auxin based on the amount of curvature induced by the unknown solution.

Procedure 20.4
Bioassay auxin

1. Working in groups of three to five students each, obtain a tray of dark-grown pea seedlings that are 8 to 10 days old.
2. Locate the third internode. Cut from the seedlings thirty-five uniform sections (one per seedling) 3 cm long from just below this internode.
3. Use a razor blade to split three-fourths of the length of each stem segment.

20-4

TABLE 20.3

EFFECTS OF GIBBERELLIC ACID ON CORN PLANTS

Plant and Treatment	Observations
Normal, untreated	
Dwarf, untreated	
Normal, treated with GA	
Dwarf, treated with GA	

TABLE 20.4

STEM CURVATURE AS A RESPONSE TO KNOWN AND UNKNOWN CONCENTRATION OF AUXIN

Petri Plate	Concentration	Log_{10} Auxin Concentration	Curvature Degrees	Mean
1	buffer alone	—	___ ___ ___ ___ ___	___
2	10 mM auxin	1	___ ___ ___ ___ ___	___
3	0.1 mM maxin	−1	___ ___ ___ ___ ___	___
4	0.01 mM auxin	−2	___ ___ ___ ___ ___	___
5	0.001 mM auxin	−3	___ ___ ___ ___ ___	___
6	0.0001 mM auxin	−4	___ ___ ___ ___ ___	___
7	unknown	_____	___ ___ ___ ___ ___	___

4. Soak the split stems in distilled water for 20 to 30 minutes to remove some of the endogenous (i.e., naturally occurring) auxin.

5. Label seven petri plates 1 to 7 to receive auxin at one of the concentrations listed in table 20.4. Add to each petri plate 20 ml of the appropriate solution listed in table 20.4. Each solution has been adjusted to pH 5.9 with 10 mM phosphate buffer.

6. Place five split stems in each of the petri plates.

7. Incubate the soaking sections at room temperature in the dark.

8. Examine the stems after 24 hours.

9. Use a protractor to measure the curvature of each stem for each treatment; record your data in table 20.4. Calculate and record the mean curvature for each treatment.

10. Plot the average curvature versus the log_{10} of the known concentrations of auxin on the graph paper at the end of this exercise.

11. Determine the unknown concentration of auxin by first locating the average curvature value of the unknown treatment on the y-axis of your standard curve (see fig. B.4). Draw a line from this point parallel to the x-axis until the line intersects the standard curve. Draw a line from this intersection straight down until it intersects the x-axis. This point on the x-axis marks the concentration of the unknown solution.

12. Record the concentration of the unknown solution in table 20.4.

Question 4

a. According to your data, what is the approximate concentration of auxin in the unknown solution?

_____

b. What is the relationship between auxin concentration and stem curvature? _____

_____

2,4-D

You will measure the effect of 2,4-D on elongation of cucumber roots. You'll then bioassay a solution of 2,4-D of unknown concentration by comparing its effect on root elongation with that induced by known concentrations of 2,4-D.

TABLE 20.5

MEAN ROOT LENGTH AS A RESPONSE TO KNOWN AND UNKNOWN CONCENTRATIONS OF 2,4-D

2,4-D Concentration	Log₁₀ Concentration of 2,4-D	Mean Root length (mm)
buffer alone	—	____
10 mg/L	1	____
1 mg/L	0	____
0.1 mg/L	-1	____
0.01 mg/L	-2	____
0.001 mg/L	-3	____
unknown	____	____

Procedure 20.5

Bioassay 2,4-D

1. Working in groups of three to five students each, label each of seven petri plates to receive 2,4-D at one of the concentrations listed in table 20.5.

2. Place several pieces of sterile filter paper into each petri plate.

3. Add to each petri plate 10 mL of the appropriate solution listed in table 20.5. Each solution has been adjusted to pH 5.9 with 10 mM phosphate buffer.

4. Place twenty cucumber seeds in each petri plate, and store each dish in the dark.

5. After 5 days, measure the length of the primary root of each seedling. Calculate and record the mean length for roots in each treatment in table 20.5.

6. Plot the mean root length versus the $\log_{10}$ of the known concentrations of 2,4-D on the graph paper at the end of this exercise.

7. Determine the unknown concentration of 2,4-D in a manner similar to that described in step 11 of procedure 20.4 for auxin bioassay. Record the concentration of the unknown solution in table 20.5.

Question 5

a. What conclusions can you draw from your data?

_____

_____

b. According to your data, what is the approximate concentration of 2,4-D in the unknown solution?

_____

_____

Gibberellic Acid

You will measure the effect of gibberellic acid on enlargement of radish cotyledons. You'll then bioassay a solution of gibberellic acid of unknown concentration by comparing its effect on cotyledon enlargement with that induced by known concentrations of gibberellic acid.

Procedure 20.6

Bioassay gibberellic acid

1. Working in groups of three to five students each, obtain a flat of dark-grown radish seedlings that are approximately 1.5 days old.

2. Label each of seven petri plates to receive gibberellic acid at one of the concentrations listed in table 20.6. Add to each petri plate 5 mL of the appropriate solution listed in table 20.6. Each solution has been adjusted to pH 5.9 with 10 mM phosphate buffer.

3. Excise the smaller cotyledon and all of the hypocotyl (the portion of seedling between the cotyledon and root) from 105 seedlings.

4. Place a pad of sterile filter paper in each petri plate.

5. Place fifteen excised cotyledons on the moistened filter paper in each petri plate. Use cotyledons of similar sizes in all treatments.

6. Place the plates in a humid environment (e.g., inside a plastic bag) in constant fluorescent light.

7. After 3 days, weigh the cotyledons to the nearest milligram and calculate the mean weight of a cotyledon in each of the treatments. Record these mean weights in table 20.6.

8. Plot the mean weight of the cotyledons versus the $\log_{10}$ of the known concentrations of gibberellic acid on the graph paper at the end of this exercise.

TABLE 20.6

MEAN WEIGHT AS A RESPONSE TO KNOWN AND UNKNOWN CONCENTRATIONS OF GIBBERELLIC ACID

Concentration of Gibberellic Acid	Log₁₀ Concentration of Gibberellic Acid	Mean Weight (mg)
buffer alone	—	___
10 mg/L	1	___
1 mg/L	0	___
0.1 mg/L	−1	___
0.01 mg/L	−2	___
0.001 mg/L	−3	___
unknown	___	___

9. Determine the unknown concentration of gibberellic acid in a manner similar to that described in step 11 of procedure 20.4 for auxin bioassay. Record the concentration of the unknown solution in table 20.6.

Question 6

a. What are your conclusions? _____

_____

b. According to your data, what is the approximate concentration of gibberellic acid in the unknown

solution? _____

Summary of Activities

1. Observe the effects of the presence and absence of auxin on *Coleus* plants.
2. Observe the effects of 2,4-D on plant growth.
3. Observe the effects of gibberellic acid on plant growth.
4. Conduct bioassay of auxin, 2,4-D, and gibberellic acid.

Questions for Further Thought and Study

1. What are some other uses for substances that regulate plant growth?
2. How do your bioassay data compare with those of other student groups? What are possible explanations for differences? What are some sources of error in a bioassay?
3. How can an increased amount of data (i.e., replicate sets) improve the certainty of your conclusions?
4. Why do your data provide only an estimation of the concentration of the unknown solutions rather than an exact concentration?
5. What techniques other than bioassay would more precisely determine the concentrations of the unknown solutions?
6. What are some potential uses of bioassays in medicine? in agriculture?
7. What are some other effects of auxin? 2,4-D? gibberellic acid?

Doing Botany Yourself

Design an experiment to bioassay an extract of leaf litter to find out whether it inhibits or promotes seed germination.

Writing to Learn Botany

Briefly describe the value of bioassays as opposed to direct chemical measurement. What assumptions must be made to validate that a bioassay for a chemical is accurate?

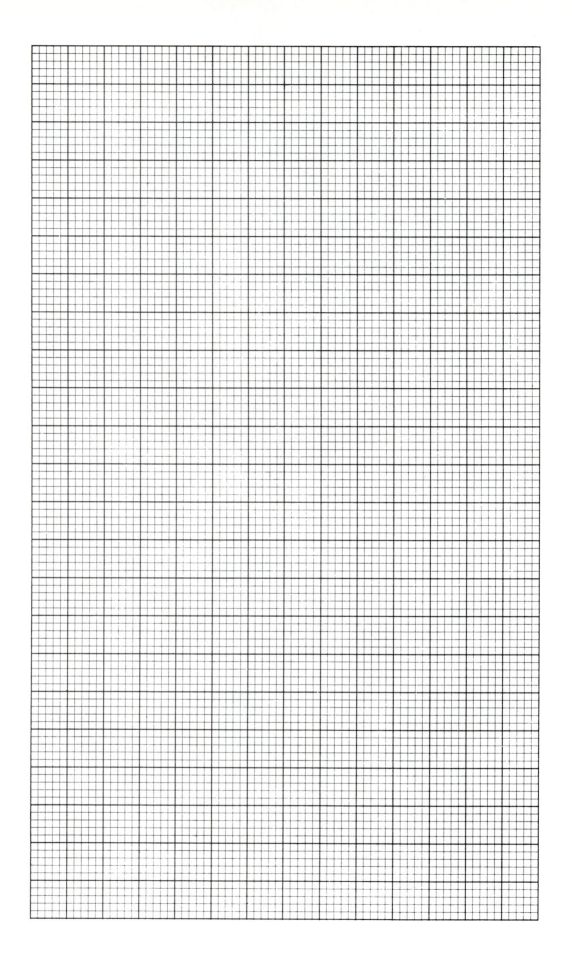

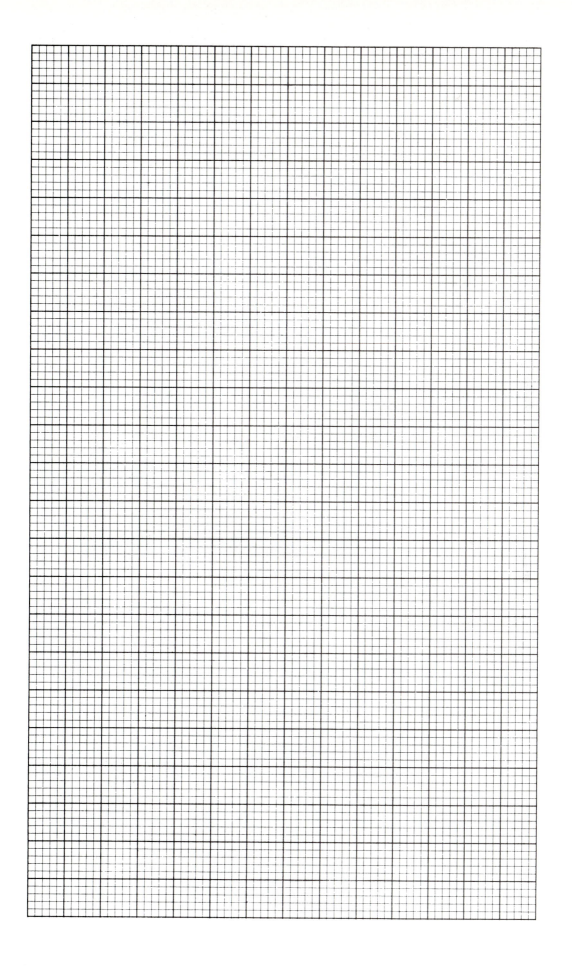

21

Plant Responses to External Factors

Objectives

By the end of this exercise you will be able to

1. describe the mechanism of phytochrome activation
2. describe the effect of light on seed germination and stem etiolation
3. describe the most common symptoms of nutrient deficiency
4. understand phototropism and gravitropism
5. describe some responses of seedlings to light and gravity.

SEED GERMINATION

Several environmental factors affect seed germination. For example, seeds of many plants germinate only in response to certain types of light. The family of pigments that absorbs light that affects seed germination (and several other developmental responses) is **phytochrome;** this pigment alternates between two forms depending on the light it has absorbed. Phytochrome is activated by the red light (660 nm) of sunlight and is inactivated by far-red light (730 nm) and/or darkness. The activation and deactivation reactions are reversible, and the ultimate physiological effect induced by the phytochrome depends on which wavelengths were absorbed last. Indeed, more than fifty different developmental processes are affected by phytochrome absorption of red and far-red light (fig. 21.1).

Grand Rapids lettuce seeds (*Lactuca sativa*) are excellent models for examining the effects of light on germination because they are sensitive to light and germinate quickly. The seeds are dormant when they are first shed, and germination is poor even with adequate oxygen and heat. Dormancy is broken by light absorption by phytochrome. The phytochrome usually exists in the red-absorbing form. Activation (absorption of red light)

causes increased water absorption by the radicle cells, and the radicle grows and elongates. This, in turn, triggers germination.

Procedure 21.1
Observe germination of lettuce seeds

1. Obtain some Grand Rapids lettuce seeds from your laboratory instructor.
2. Working in small groups, place fifty seeds in each of six petri plates containing water-soaked filter paper.
3. Cover the petri plates with lids and expose the seeds to the following treatments:

   Treatment 1: Continuous darkness (wrap the plate in metal foil)

   Treatment 2: Continuous light

   Treatment 3: Red light for 10 minutes, then darkness

   Treatment 4: Far-red light for 10 minutes, then darkness

   Treatment 5: Red light for 10 minutes, followed by far-red light for 10 minutes, then darkness

   Treatment 6: Red light for 10 minutes, far-red light for 10 minutes, red light for 10 minutes, then darkness

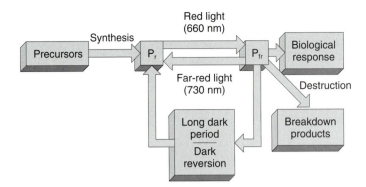

Red light
(660 nm)

Synthesis

Precursors → P_r ⇄ P_{fr} → Biological response

Far-red light
(730 nm)

Destruction

Long dark period
Dark reversion

Breakdown products

FIGURE 21.1

Phytochrome absorbs light and affects a variety of plant processes, including flowering. Phytochrome is synthesized as P_r, which absorbs red light and is converted to P_{fr}. P_{fr} has several possible fates; it can absorb far-red light and be converted back to P_r, initiate a biological response, be destroyed, or revert to P_r if maintained in darkness.

TABLE 21.1

GERMINATION OF LETTUCE SEEDS UNDER SIX LIGHT TREATMENTS

Treatment	Percent Germinated
1. Darkness	___
2. Continuous light	___
3. Red light, then darkness	___
4. Far-red light, then darkness	___
5. Red light, followed by far-red light, then darkness	___
6. Red light, far-red light, red light, then darkness	___

4. During your next lab period, determine the percentage of germination resulting from each treatment.

5. Record your results in table 21.1.

Question 1

a. What do you conclude about the influence of light on germination of lettuce seeds? _____

_____

Germinating seeds of different plants often respond differently to light. Germination of onion seeds, for example, is affected by light versus dark. Determine their sensitivity with procedure 21.2.

Procedure 21.2
Observe germination of onion seeds

1. Divide fifty onion seeds; placing twenty-five in each of two petri plates containing water-soaked filter paper.

2. Put lids on the petri plates and label the lids "Light" and "Dark." Place one petri plate in room light and the other plate in darkness.

TABLE 21.2

GERMINATION OF ONION SEEDS IN LIGHT AND DARK

Treatment	Percent Germinated
Room light	___
Darkness	___

3. Examine the seeds during your next lab period and record the percentage germination in table 21.2.

Question 2

a. Does light promote or inhibit germination of onion seeds? _____

b. How is this response different from that of lettuce?

_____

_____

TABLE 21.3

PLANT GROWTH UNDER DIFFERENT LIGHT TREATMENTS

Treatment	Leaf Size	Stem Diameter	Height	Color of Shoots	Stem Strength
Light					
Darkness					
Light, then darkness					
Darkness, then light					

PLANT GROWTH AFFECTED BY LIGHT

Many seeds germinate in the dark and push their stems above the soil to absorb light. If the seedlings do not absorb light, their leaves will be small and the plant will appear pale and spindly. Such a plant is **etiolated.** Chlorophyll does not develop until the plant is exposed to light. Activation of phytochrome helps promote normal growth. Use procedure 21.3 to observe etiolation.

Procedure 21.3
Examine etiolation

1. Obtain two groups of bean (*Phaseolus*) seedlings that are 10 to 14 days old. One group was grown in the dark and one group was grown in light.
2. Complete the first two rows of table 21.3 with your observations.
3. After completing and recording your observations, reverse the two treatments and examine the plants during your next lab period.
4. Complete the last two rows of the table with your final observations.

Question 3
a. How did the seedlings manage to grow at all while in the dark? _____

_____

b. Did etiolation include stem elongation? _____
c. What is the advantage of rapid stem elongation of a seed germinating in the dark? _____

_____

PLANT NUTRITION

Growth of green plants requires suitable temperature and adequate amounts of carbon dioxide, oxygen, water, light, and a group of essential elements. Carbon, hydrogen, and oxygen comprise 98% of the fresh weight of plants. The remaining 2% consists of thirteen other elements classified either as macronutrients or micronutrients.

- **Macronutrients** are nutrients needed in relatively large amounts (100–2,000 ppm). Nitrogen (N), phosphorus (P), calcium (Ca), potassium (K), magnesium (Mg), and sulfur (S) are examples of macronutrients.

- **Micronutrients** are nutrients needed in relatively small amounts (<100 ppm). Iron (Fe), chlorine (Cl), copper (Cu), manganese (Mn), zinc (Zn), molybdenum (Mo), and boron (B) are examples of micronutrients.

Although plants need more macronutrients than micronutrients, all are essential for plant growth. The absence of any essential micronutrient or macronutrient ultimately kills a plant.

Before your lab period, seeds were germinated in distilled water and washed sand. Then the young seedlings were transferred to a variety of growth media. Control plants were watered with a solution containing all essential nutrients. Hoagland's solution is a common mixture of nutrients used to provide for healthy growth (table 21.4). Your control plants were watered with a solution similar to Hoagland's. Other groups of seedlings were watered with solutions that were deficient in an essential element. Follow procedure 21.4 to observe the effects of nutrient deficiency.

TABLE 21.4

HOAGLAND'S SOLUTION, A COMMON NUTRIENT MEDIUM FOR HEALTHY PLANT GROWTH

Macronutrients	Grams/liter	Micronutrients	Grams/liter
$Ca(NO_3)_2 \cdot 4H_2O$	1.18	H_3BO_3	0.60
KNO_3	0.51	$MnCl_2 \cdot 4H_2O$	0.40
$MgSO_4 \cdot 7H_2O$	0.49	$ZnSO_4$	0.05
KH_2PO_4	0.14	$CuSO_4 \cdot 5H_2O$	0.05
		$H_2Mo_4 \cdot 4H_2O$	0.02
		ferric tartrate	0.50

TABLE 21.5

SYMPTOMS OF NUTRIENT DEFICIENCY

Nutrient Solution	Symptom(s)
Complete (containing all required minerals)	
Complete solution minus Ca	
Complete solution minus N	
Complete solution minus P	
Complete solution minus Mg	
Complete solution minus K	
Complete solution minus S	
Complete solution minus Fe	

Procedure 21.4

Examine plants with symptoms of nutrient deficiency

1. Examine the plants on demonstration that have been grown in nutrient solutions lacking Ca, N, P, Mg, K, S, and Fe.

2. Record the symptoms for deficiency of each nutrient in table 21.5. Do not refer to table 21.6 until you have completed your observations.

3. After examining all the seedlings, refer to table 21.6 to see which deficiencies are most obvious and easily diagnosed.

21-4

TABLE 21.6

FUNCTIONS AND DEFICIENCY SYMPTOMS OF SOME MAJOR ELEMENTS

Element	Major Functions	Deficiency Symptoms
Nitrogen (N)	Major component of amides, amino acids, and proteins. Present in membranes, organelles, and the cell wall. Balance of carbohydrates and nitrogenous substances necessary.	Leaves often more erect. Stem and leaves stunted with excess root development. Foliage, especially older leaves, chlorotic. Unable to flower.
Phosphorus (P)	Constituent of phospholipids, nucleic acids, and nucleoproteins. Important in respiration and energy transfer.	Small plants, narrow leaves, root system larger but with fewer laterals. Accumulation of sugars in older leaves promotes the synthesis of purple anthocyanin pigments. Stiff but weak stems and leaves. Older leaves yellowed. Other leaves dark green.
Potassium (K)	Not known to be structurally part of organic compounds. Role is likely catalytic and regulatory. Needed to activate several enzyme systems.	Internodes short, stems weak. Localized chlorotic or molting of older leaves, particularly at the tips and margins. Later stages may have necrotic mottling. Leaf margins frequently curled under.
Magnesium (Mg)	Constituent of chlorophyll. Important cofactor for enzymes in respiration and in phosphate metabolism.	Older leaves become chlorotic between the veins at the tips and margins. Usually not characterized by necrotic spots. Root system frequently overdeveloped. Leaf margins may cup upward.
Calcium (Ca)	Component in pectin compounds of middle lamella. Present in organic acids bound to proteins. Plays a role in nitrogen metabolism and membrane integrity.	Deficiency may cause iron uptake imbalance particularly with magnesium. Young leaves are affected first. Tips and margins of leaves become light green and later necrotic. Tips of leaves become limp. Terminal bud often dies.
Sulfur (S)	Component of proteins. Component of iron-sulfur proteins of the electron transport system.	Younger leaves light green. Veins lighter than intervein area.
Iron (Fe)	Electrons transported in cytochromes.	Effects localized on new leaves. Leaves chlorotic. Veins remain green.

Question 4

a. How would you "cure" a plant suffering from these deficiency symptoms? _____

_____

Commercial Fertilizers

Commercial fertilizer contains three primary nutrients: nitrogen, phosphorus, and potassium. It may also contain other macronutrients and/or micronutrients.

Examine the bag of fertilizer in the lab. The contents of the bag are described by three numbers on the bag's label. For example, a bag labeled 10-12-8 is 10% nitrogen (as ammonium or nitrate), 12% phosphorus (as phosphoric acid), and 8% potassium (as mineral potash).

PLANT TROPISMS

A **tropism** is a movement in response to an external stimulus. The direction of the movement is determined by the direction of the most intense stimulus. Two tropisms you will study in this lab are **phototropism,** which is directed growth in response to light, and **gravitropism,** or growth in response to gravity (see plate IX, fig. 16, and plate XIII, fig. 33).

TABLE 21.7

OBSERVATIONS OF PHOTOTROPISM

Time (h)	Mean Curvature (degrees)
0.5	_____
1.0	_____
1.5	_____
2.0	_____

Phototropism

Procedure 21.5
Observe phototropism

1. Obtain from your lab instructor some 10- to 14-day-old radish (*Raphanus*) seedlings. These seeds have been grown in diffuse, overhead light.

2. At the beginning of the lab, place your seedlings approximately 25 cm from a 100-watt light so that light strikes the shoots at a right angle.

3. Use a protractor to measure curvature of the seedlings every 30 minutes for 2 hours.

4. Record your results in table 21.7.

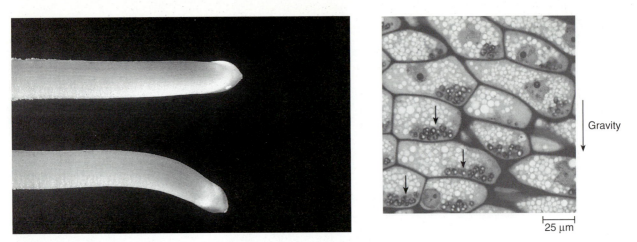

FIGURE 21.2

Perception of gravity in root caps. Cells in the center of a root cap contain numerous starch-laden amyloplasts located in the lower part of the cells (*right photo*). Amyloplasts (*arrows*) sediment to the bottom of the cell when roots are oriented horizontally. This sedimentation of amyloplasts has long been thought to be the basis for how roots perceive gravity (*left photo*).

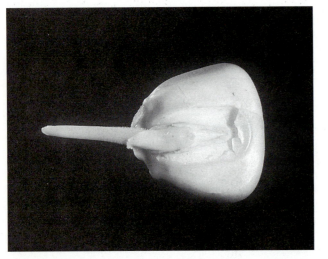 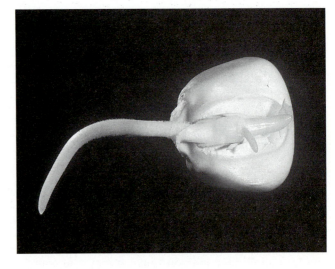

FIGURE 21.3

Gravitropism by horizontally oriented roots of corn (*Zea mays*). Downward curvature begins within 30 minutes and is completed within a few hours. Curvature results from faster elongation of the upper side of the root than of the lower side.

Question 5

a. Which direction did the seedlings curve? _____

b. Does this curvature indicate positive or negative phototropism? _____

c. What is the adaptive significance of phototropism?

_____

_____

d. Would you expect roots to react similarly to light?

_____

Why or why not? _____

_____

Gravitropism

Plants may perceive gravity by the movement of starch-laden amyloplasts within cells (fig. 21.2). Growing corn roots show their response to gravity by curving down (fig. 21.3).

Yesterday, corn seedlings with roots approximately 1 cm long were placed in a glass beaker. Seedlings labeled "H" had their root oriented horizontally, whereas those labeled "D" had their root pointing down. The terminal 3 to 4 mm was removed from roots of seedlings labeled with an asterisk (*).

TABLE 21.8

OBSERVATIONS OF GRAVITROPISM IN ROOTS

Treatment or Orientation	Direction of Growth
Intact roots oriented vertically (V)	＿＿
Intact roots oriented horizontally (H)	＿＿
Detipped roots oriented vertically (V*)	＿＿
Detipped roots oriented horizontally (H*)	＿＿

TABLE 21.9

OBSERVATIONS OF GRAVITROPISM IN STEMS

Treatment/Time	Distance of Stem Tip from Table Surface
Horizontal 0 minutes	＿＿
Horizontal 30 minutes	＿＿
Horizontal 60 minutes	＿＿
Horizontal 90 minutes	＿＿
Horizontal 120 minutes	＿＿
Observation	
Inverted, 24 hours	
Horizontal, 24 hours	

Procedure 21.6
Observe root gravitropism

1. Obtain the containers of corn seedlings oriented horizontally and vertically.
2. Examine the experimental setup and the direction of root growth.
3. Record your observations in table 21.8.

Question 6
a. Which roots grew down? ＿＿＿＿＿＿＿＿＿＿＿＿

＿＿＿＿＿＿＿＿＿＿＿＿＿＿＿＿＿＿＿＿＿

b. Which one(s) didn't? ＿＿＿＿＿＿＿＿＿＿＿＿

＿＿＿＿＿＿＿＿＿＿＿＿＿＿＿＿＿＿＿＿＿

c. What conclusion do you draw from these data about

the part of the root that perceives gravity? ＿＿＿＿＿

＿＿＿＿＿＿＿＿＿＿＿＿＿＿＿＿＿＿＿＿＿

d. Where in roots does the differential growth occur

that produces gravitropism? ＿＿＿＿＿＿＿＿＿

＿＿＿＿＿＿＿＿＿＿＿＿＿＿＿＿＿＿＿＿＿

e. Are roots positively or negatively gravitropic?

＿＿＿＿＿＿＿＿＿＿＿＿＿＿＿＿＿＿＿＿＿＿＿

f. What is the adaptive significance of gravitropism?

＿＿＿＿＿＿＿＿＿＿＿＿＿＿＿＿＿＿＿＿＿＿＿

＿＿＿＿＿＿＿＿＿＿＿＿＿＿＿＿＿＿＿＿＿＿＿

Procedure 21.7
Examine stem gravitropism

1. Obtain containers of tomato plants (*Lycopersicon*) or sunflower plants.
2. Turn two or three of the potted plants horizontally.
3. Measure the distance from the stem tip to the table's surface. Record your measurements in table 21.9.
4. Every 30 minutes remeasure the distance between the stem tips and table, and record your results.
5. Some of the plants available in the lab were turned horizontally and inverted yesterday; others were left upright.
6. Examine stem curvature after 24 hours, and record your observations in table 21.9.

Question 7

a. How are the stems oriented after 24 hours? _____

_____

b. What is the adaptive advantage of a gravitropic
    response? _____

_____

Summary of Activities

1. Observe the effect of light on lettuce and onion seed
    germination.
2. Examine etiolation of bean seedlings.
3. Examine symptoms of nutrient deficiency.
4. Observe phototropism in radish seedlings.
5. Observe gravitropism in corn seedlings.
6. Observe gravitropism in tomato plants.

Questions for Further Thought and Study

1. How are phototropism and gravitropism similar?
    How are they different?
2. What is the adaptive significance of the activation of
    phytochrome by far-red light?
3. What is the adaptive significance of seeds not
    germinating in darkness?

Doing Botany Yourself

Design an experiment to determine if seed
germination by tropical species of plants is more sensitive to light than that of temperate species.

Writing to Learn Botany

Describe the characteristics of plants you
would choose to grow in outer space. Base your choices
on nutrient requirements, phototropic sensitivity, and
gravitropic sensitivity.

Transpiration

Objectives

By the end of this exercise you will be able to

1. describe the structure and function of stomata
2. discuss how environmental conditions such as wind and light affect stomatal opening
3. measure the transpiration rate of a plant.

To survive and reproduce, land plants must cope with many environmental problems. Among the biggest of these problems are obtaining water and avoiding desiccation (water loss). "Solutions" to this and other challenges require expenditure of energy and an efficient structure and physiology. Structural and functional solutions to problems such as maintaining water balance are called adaptations, and the study of these adaptations is among the most interesting aspects of biology. In this exercise you will study the "water physiology" of plants—how plants are adapted to control their water content and evaporative loss. Before starting today's exercise, review in your textbook the structure and function of xylem and stomata, and review how water moves through plants (see plate VII, fig. 12).

    The loss of water from plants is called transpiration. Most transpiration occurs through stomata of leaves. However, transpiration also occurs in flowers. Demonstrate this with the following procedure.

Procedure 22.1
Visualize transpiration in flowers

1. At the beginning of the laboratory period obtain a small beaker containing methylene blue from your lab instructor.

> ### C A U T I O N
> Be careful not to spill methylene blue. It will stain your skin and clothes!

2. Working in small groups, submerge a stem of periwinkle (*Catharanthus*) containing a white flower so that only about 0.5 cm of the stem extends below the flower into the dye. Cut the submerged stem.
3. Float the flower in the dye solution so that the cut portion of the stem is submerged in the dye.
4. Set the beaker containing the flower and dye in an illuminated area, and examine the flower after 2 to 3 hours.

Question 1
a. In which part of the flower is the dye located? _____
b. What do you conclude from this observation? _____
_____

PLANT-WATER RELATIONS

Leaves of most plants have epidermal pores, which are formed by the separation of a pair of **guard cells** (fig. 22.1). Together, the guard cells and the pore between them compose a **stoma** (pl., stomata). In a previous exercise you learned that CO_2 is essential for photosynthesis and reaches the photosynthetic cells in the leaf via stomata. If enough water is available, guard cells absorb water and become turgid. When guard cells are turgid they expand and bend to form a **stomatal pore** through which CO_2 enters and water vapor exits a leaf. When water is scarce, guard cells lose their turgidity and shrink. As a result, guard cells touch each other and the stomatal pore closes. Thus, the

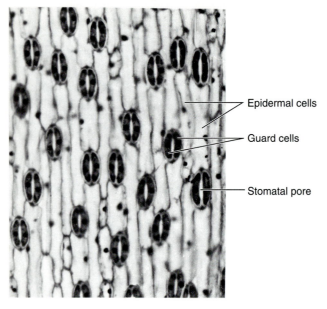

FIGURE 22.1

Stomata on the surface of a lily (*Lilium*) leaf.

4. Examine the tissue with your microscope and diagram the epidermis and its stomata (in the space following).

5. Make a wet mount of epidermis from both sides of the leaf.

6. Place a few drops of 10% NaCl at one edge of the coverslip covering the *Kalanchoë*.

7. Wick this solution under the coverslip by touching the solution at the opposite edge of the coverslip with a piece of paper towel.

8. Reexamine the stomata.

9. Observe the demonstration of an epidermal peel of *Zea mays* (corn), a monocot. In the space below, diagram cells of the epidermis of corn. Compare this diagram with your drawing from step 4.

Epidermis and Stomata   Epidermis and Stomata
    in *Kalanchoë*   in *Zea mays*

water status of a plant determines whether stomata are open or closed and indirectly determines whether or not gases move in and out of the leaf.

STOMATAL STRUCTURE

Guard cells in dicots are bean-shaped and attached to each other at their ends. Microfibrils form transverse bands around the cell so that when water enters the cells they cannot expand around the middle. Instead, they lengthen and bow apart, thereby opening the stomatal pore. Use procedure 22.2 to examine living guard cells and how they respond to changing environmental conditions.

Procedure 22.2
Observe stomatal structure

1. Obtain a leaf from a well-watered *Kalanchoë* plant that has been illuminated for 4 to 5 hours.

2. Bend the leaf until it snaps; then pull the surface portion of the leaf away from the rest of the leaf in a slow downward motion (your lab instructor will demonstrate this technique). A thin sheet of transparent epidermis will tear away from the leaf.

3. Quickly place the piece of epidermis in a few drops of distilled water on a microscope slide (do not let the tissue desiccate), and place a coverslip on the tissue. Be careful not to trap any air bubbles beneath the tissue or coverslip.

Question 2

a. Approximately how many stomata are in a mm² of the lower surface area of *Kalanchoë*? _____

b. Are the densities of stomata similar on the upper and lower surfaces of the *Kalanchoë* leaf? _____

c. In well-watered plants, the guard cells of *Kalanchoë* are turgid and form a stomatal pore. Why? _____

_____

_____

d. Are stomatal densities similar in *Zea* and *Kalanchoë*?

_____

e. Are stomata of the two plants arranged in a pattern, or do they occur randomly? _____

f. How is the structure of individual stoma different in *Zea* than in the *Kalanchoë*? _____

_____

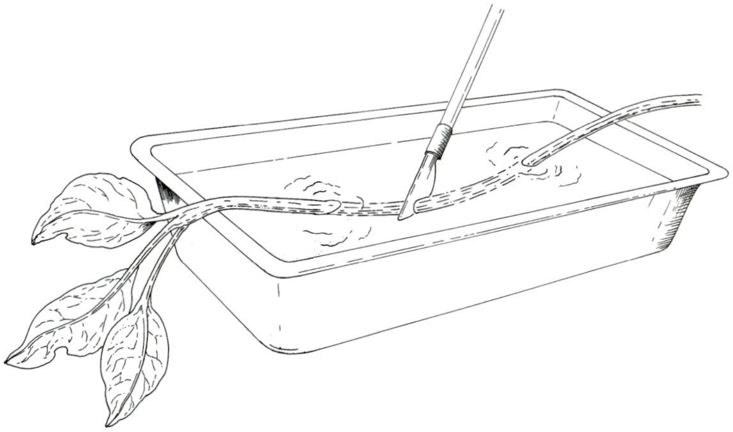

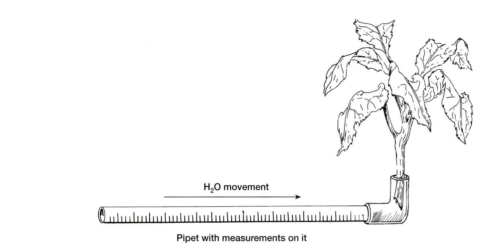

FIGURE 22.2

Method of cutting submerged stem.

H₂O movement

Pipet with measurements on it

FIGURE 22.3

Experimental setup for measuring transpiration.

g. Did the NaCl solution cause stomata to close? _____ Why or why not? _____

_____

h. What is the advantage of closed stomata when water is in short supply? _____

_____

i. What are the disadvantages? _____

_____

TRANSPIRATION

Transpiration is the loss of water from plants; it is strongly influenced by environmental conditions such as CO_2, light, wind, and humidity. Use procedure 22.3 to measure the effects of light and wind on transpiration.

Procedure 22.3
Quantify transpiration

1. Obtain a sunflower (*Helianthus*) plant from your lab instructor.
2. Submerge the stem in water and obliquely cut the stem with a sharp scalpel (fig. 22.2).
3. While it is still submerged in water, attach the severed stem to a water-filled tube connected to a 1.0 mL pipet as shown in figure 22.3. Be sure that no air bubbles are in the system and that no water is on the leaves.
4. Place the stem in an illuminated area. Transpiration will move water through the pipet.
5. Allow about 10 minutes for the rate of water movement in the pipet to stabilize. Note the position of the meniscus in the pipet.
6. Allow transpiration to occur for 15 minutes.

TABLE 22.1

TRANSPIRATION RATES UNDER DIFFERENT CONDITIONS

Condition	Transpiration Rate mL H_2O h^{-1}
Control (standard illumination)	_____
Darkness or reduced light	_____
Light; mild breeze (made by a small fan)	_____
Dark; mild breeze	_____
Light; leaves coated with petroleum jelly	_____
Light; all leaves removed	_____

7. Determine and record the volume of water (mL) transpired by observing the distance that the meniscus moved.

8. Multiply this value by 4 to convert to milliliters per hour. Record this rate of transpiration in table 22.1 as the control value.

9. Continue the experiment, and determine the transpiration rate under the conditions in the order listed in table 22.1.

10. If data from other groups are available, determine the mean transpiration rates for each condition and record them next to your group's values in table 22.1.

Question 3

a. State your conclusions about the influence of light, darkness, moving air, clogged stomata, and removed leaves on water movement in plants.

Light: _____

Dark: _____

Breeze: _____

Clogged stomata: _____

Removed leaves: _____

b. How might the number and size of leaves affect the

transpiration rate of a plant? _____

_____

c. Would you expect leaves of desert plants to have

more or fewer stomata than tropical plants? _____

Why? _____

_____

Summary of Activities

1. Observe transpiration in flowers.
2. Make a wet mount of *Kalanchoë* and examine stomatal structure.
3. Examine the response of stomata to a salt solution.
4. Quantify rates of transpiration under various conditions.

Questions for Further Thought and Study

1. Why was it important in this lab exercise to sever the stems while they were submerged in liquid?

2. What are some structural features of plants that function to minimize water loss?

3. What plant adaptations would improve water economy in plants? Which plants are closest to illustrating those adaptations?

Doing Botany Yourself

Assume that a maple tree has 100,000 leaves, each having an average area of 35 cm^2. If the transpiration rate of each leaf is 0.05 ml H_2O h^{-1} cm^{-2} leaf area, how many liters of water move through the plant in a day?

Writing to Learn Botany

Describe any adaptations that would improve water economy in plants. In which environments might you expect to find the adaptations you described?

Evolution

23

Objectives

By the end of this exercise you will be able to

1. demonstrate the effects of selection pressure on genotypic and phenotypic frequencies in a population

2. understand the significance of the Volvocine line of evolution

3. understand how a mutation affecting the plane of cellular division could result in the evolution of morphologically different body forms.

Evolution broadly describes genetic change in populations. The existence of genetic change (and therefore evolution) is universally accepted. We know that many mechanisms can change the genetic makeup of populations, but the relative importance of each mechanism is controversial. Stochastic and random events such as **mutations** (changes in the genetic message of a cell) and catastrophes (e.g., meteor showers, ice ages) are all responsible to some degree for genetic change. However, Charles Darwin and others formulated a theory that many biologists believe explains the major force behind genetic change.

Darwin postulated that organisms that survive and reproduce successfully have genetic traits aiding survival and reproduction. These traits enhance an organism's **fitness,** which is its tendency to produce more offspring than competing individuals produce. Because the most fit organisms transmit their traits to their offspring, traits promoting survival and reproduction are selectively passed to the next generation. After many generations the frequency of these traits increases in the population, and the nature of the population gradually changes. Darwin called this process **natural selection** and proposed the process as a major force guiding genetic change and the formation of new species. Review in your textbook the theories of evolution and the mechanism of natural selection (see plate XVI, fig. 45).

Showing the effects of natural selection in living populations is time-consuming and tedious. In this exercise you will simulate reproducing populations with colored beads representing plants and their gametes. With this artificial population you can quickly follow genetic change over many generations. Before you begin work, read exercise 10, "Genetics," especially the terms **gene, allele, dominant alleles, recessive alleles, homozygous,** and **heterozygous.**

You will begin your experiments using a "stock population" of plants consisting of a container of colored beads. Each bead represents a haploid gamete, and the color of the "gamete" (red or white) represents the allele it is carrying. Individual organisms from this model population are diploid and therefore are represented by two beads.

UNDERSTANDING ALLELIC AND GENOTYPIC FREQUENCIES

Frequency is the proportion of individuals in a certain category relative to the total number of individuals in all categories. The frequency of an allele or genotype is expressed as a proportion of the total alleles or genotypes in a population. For example, if one-fourth of the individuals of a population are genotype Rr, then the frequency of Rr is 0.25. If three-fourths of all alleles in a population are R, then the frequency of R is 0.75.

In this exercise you will simulate evolutionary changes in allelic and genotypic frequencies in an artificial population. The trait you will work with is flower color. A red bead is a gamete with a dominant allele (complete dominance) for red flowers (R), and a white

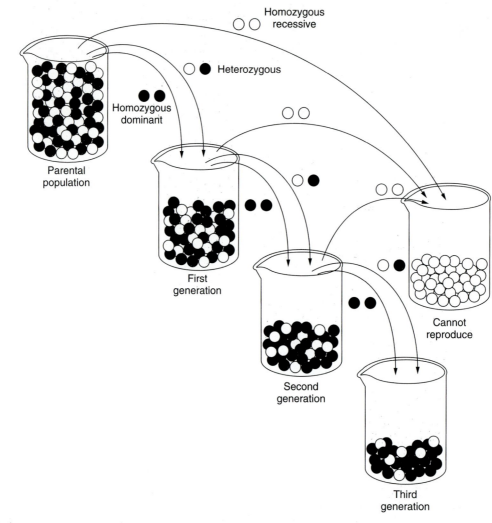

FIGURE 23.1

Demonstration of the effect of 100% selection pressure on genotypic and phenotypic frequencies across three generations. Progressive generations are produced from a parental population, and selection is against the homozygous recessive genotype. Random mating within the parental population is simulated by mixing the gametes (beads), and the parental population is sampled by removing two alleles (i.e., one individual) and placing them in the next generation. Homozygous recessive individuals are removed (selected against) from the population. The genotypic and phenotypic frequencies after production of each generation are recorded. The production of each generation depletes the beads in the previous generation in this simulation.

bead is a gamete with a recessive allele for white flowers (*r*). An individual is represented by two gametes (beads). Individuals with genotype *RR* have red flowers and those with *rr* have white flowers (fig. 23.1).

Procedure 23.1
Establish a parental population

1. Obtain a "stock population" of organisms consisting of a container of red and white beads.

2. Obtain an empty container marked "Parental Population."

3. From the stock population select twenty-five homozygous dominant individuals (*RR*) and place

them in the container marked "Parental Population." Each individual is represented by two red beads.

4. From the stock population select fifty heterozygous individuals (*Rr*) and place them in the container marked "Parental Population." Each individual is represented by a red and a white bead.

5. From the stock population select twenty-five homozygous recessive individuals (*rr*) and place them in the container marked "Parental Population." Each individual is represented by two white beads.

6. Record in table 23.1 the correct genotypic frequencies for your newly established parental population.

7. Complete table 23.1 with the number and frequency of each of the two alleles.

TABLE 23.1

FREQUENCIES OF GENOTYPES AND ALLELES OF THE PARENTAL POPULATION

Genotypes	Frequency	Alleles	Frequency
RR	_____	R	_____
Rr	_____	r	_____
rr	0.25		

Question 1

a. How many of the total beads are red and how many are white? _____

b. What color flowers do *Rr* individuals have? _____

c. How many beads represent the population? _____

THE HARDY-WEINBERG LAW

The **Hardy-Weinberg Law** is a statement with math equations that enable us to calculate and predict allelic and genotypic frequencies. We can compare these predictions with actual changes that we observe in natural populations and learn about factors that influence gene frequencies.

This predictive model includes two simple equations first described for stable populations by G. H. Hardy and W. Weinberg. Hardy-Weinberg equations (1) predict allelic and genotypic frequencies based on data for only one or two frequencies and (2) provide a set of theoretical frequencies that we can compare to frequencies from natural populations. For example, if we know the frequency of *R* or *RR*, we can calculate the frequency of *r*, *Rr*, and *rr*. Then we can compare these frequencies with those of a natural population that we might be studying. If we find variation from our predictions we can study the reasons for this genetic change. This comparison is important because biological characteristics of natural populations rarely correspond exactly to theoretical calculations. Furthermore, deviations are important because they often reveal to scientists unknown factors influencing the population being studied.

According to the Hardy-Weinberg Law, the frequency of the dominant allele of a pair is represented by the letter *p* and that of the recessive allele by the letter *q*. The genotypic frequencies of *RR* (homozygous dominant), *Rr* (heterozygous), and *rr* (homozygous recessive) are represented by p^2, $2pq$, and q^2, respectively. Examine the frequencies in table 23.1 and verify the Hardy-Weinberg equations:

$$p + q = 1$$
$$p^2 + 2pq + q^2 = 1$$

The Hardy-Weinberg Law and its equations predict that frequencies of alleles and genotypes will remain constant from generation to generation in stable populations. Therefore, these equations can be used to predict genetic frequencies through time. However, the Hardy-Weinberg prediction assumes that

1. the population is large enough to overcome random events;
2. pairing for sexual reproduction is random;
3. mutation does not occur;
4. individuals do not migrate into or out of the population;
5. there is no selection pressure.

Question 2

a. Consider the Hardy-Weinberg equations. If the frequency of a recessive allele is 0.3, what is the frequency of the dominant allele? _____

b. If the frequency of the homozygous dominant genotype is 0.49, what is the frequency of the dominant allele? _____

c. If the frequency of the homozygous dominant genotype is 0.49, what is the frequency of the homozygous recessive genotype? _____

Use procedure 23.2 to verify the predictions of the Hardy-Weinberg Law. Produce a generation of offspring from the parental population that you created in the previous procedure.

Procedure 23.2
Verify the Hardy-Weinberg Law

1. Obtain the parental population established in procedure 23.1.
2. Simulate the random mating of individuals by mixing the population. Then reach into the parental container without looking and randomly select two gametes. This pair of gametes with red or white alleles represents an individual offspring.
3. Record the occurrence of the genotype in table 23.2 as a hash mark under the heading "Number of Offspring," and return the beads to the container.
4. Repeat steps 2 and 3 100 times to simulate the production of 100 offspring.
5. Calculate the frequency of each genotype and allele, and record them in table 23.2.

Question 3

a. The Hardy-Weinberg Law predicts that genotypic frequencies of offspring will be the same as those of the parental generation. Were they the same in your simulation? _____

TABLE 23.2

FREQUENCIES OF GENOTYPES AND ALLELES OF OFFSPRING FROM THE PARENTAL POPULATION

Genotype	Number of Offspring (Total = 100)	Frequency
RR	_____	_____
Rr	_____	_____
rr	_____	
		$R =$ ___
		$r =$ ___

TABLE 23.3

GENOTYPIC FREQUENCIES FOR 100% NEGATIVE SELECTION

Genotype	Generation				
	First	Second	Third	Fourth	Fifth
RR	___	___	___	___	___
Rr	___	___	___	___	___
rr	___	___	___	___	___

Note: Total for each generation = 1.0.

b. If the frequencies were different, then one of the assumptions of the Hardy-Weinberg Law was probably violated. Which one? _____

EFFECT OF A SELECTION PRESSURE

Selection is the differential reproduction of genotypes; that is, some genotypes are passed to the next generation more often than other genotypes. In positive selection, genotypes with adaptive traits in an environment increase in frequency because their bearers contribute more offspring to the next generation. In negative selection, genotypes with nonadaptive traits in an environment decrease in frequency because their bearers contribute fewer offspring.

Selection pressures are factors that affect organisms, such as temperature and predation, and result in selective reproduction of genotypes. Some pressures may elicit 100% negative selection against a characteristic and eliminate any successful reproduction by individuals with that characteristic. For example, plants with white flowers may be unseen by pollinators attracted only to bright colors. Pollinators that are attracted only to red are a negative selection pressure against white flowers. If reproduction by plants with white flowers was eliminated (i.e., there is 100% negative selection), would the frequency of white flowers in the population decrease with subsequent generations? Use procedure 23.3 to test this.

Procedure 23.3
Examine 100% negative selection pressure

1. Establish the same parental population that you used to test the Hardy-Weinberg prediction.
2. Simulate the production of an offspring from this population by randomly withdrawing two gametes to represent an individual offspring.

3. If the offspring is RR or Rr, place it in a container for production of the "Next Generation." Record the occurrence of this genotype on a separate sheet of paper.
4. If the offspring is rr, place this individual in a container for those that "Cannot Reproduce." Individuals in this container should not be used to produce subsequent generations. Record the occurrence of this genotype on a sheet of paper.
5. Repeat steps 2 to 4 until the parental population is depleted, thus completing the first generation.
6. Calculate the frequencies of each of the three genotypes that you recorded on the separate sheet, and record these frequencies for the first generation in table 23.3.
7. Repeat steps 2 to 5 to produce a second, third, fourth, and fifth generation. Individuals in the beaker labeled "Next Generation" will serve as the parental population for each subsequent generation. After the production of each generation, record your results in table 23.3.

Because some members of each generation cannot reproduce (i.e., the rr that you removed), the number of offspring from each successive generation of your population will decrease. However, the frequency of each genotype, not the number of offspring, is the most important value.

Question 4
a. Did the frequency of white individuals decrease with successive generations? _____

b. Was the decrease of white individuals from the first to second generation the same as the decrease from the second to the third generation? _____

from the third to the fourth generation? _____

Why or why not? _____

c. How many generations would be necessary to eliminate the allele for white flowers? _____

23-4

Most naturally occurring selective pressures do not eliminate reproduction by the affected individuals. Instead, their reproductive capacity is reduced by a small proportion. To show this, repeat the previous experiment, but eliminate only 20% of the *rr* offspring from the reproducing population.

Procedure 23.4
Examine 20% negative selection pressure

1. Establish the same parental population that you used to test the Hardy-Weinberg prediction.

2. Simulate the production of an offspring from this population by randomly withdrawing two gametes to represent an individual offspring.

3. If the offspring is *RR* or *Rr*, place it in a container for production of the "Next Generation." Record the occurrence of this genotype on a separate sheet of paper.

4. If the offspring is *rr*, place every fifth individual (20%) in a separate container for those that "Cannot Reproduce." Individuals in this container should not be used to produce subsequent generations. Place the other 80% of the homozygous recessives in the container for production of the "Next Generation." Record the occurrence of this genotype on a sheet of paper.

5. Repeat steps 2 to 4 until the parental population is depleted, thus completing the first generation.

6. Calculate the frequencies of each of the three genotypes that you recorded on the separate sheet and record these frequencies for the first generation in table 23.4.

7. Repeat steps 2 to 5 to produce a second, third, fourth, and fifth generation. Individuals in the "Next Generation" will serve as the parental population for each subsequent generation. After the production of each generation, record your results in table 23.4.

Because some members of each generation cannot reproduce, the number of offspring from each generation of your population will decrease. However, the frequency of each genotype, not the number of offspring, is the most important value.

Question 5
a. Did the frequency of white individuals decrease with successive generations? _____

b. Was the rate of decrease for 20% negative selection similar to the rate for 100% negative selection?

_____

If not, how did the rates differ? _____

_____

TABLE 23.4					

GENOTYPIC FREQUENCIES FOR 20% NEGATIVE SELECTION

Genotype	Generation				
	First	Second	Third	Fourth	Fifth
RR	___	___	___	___	___
Rr	___	___	___	___	___
rr	___	___	___	___	___

Note: Total for each generation = 1.0.

AN EXAMPLE OF EVOLUTION: THE VOLVOCINE LINE

Chlamydomonas and *Volvox* are green algae (division Chlorophyta) often cited as the "beginning" and "end," respectively, of a line of evolutionarily specialized algal genera known as the **Volvocine line.** Evolutionary lines are related organisms showing progressive changes in a trait. In this portion of the lab, you will study the Volvocine line of evolution, beginning with a review of unicellular *Chlamydomonas* and ending with colonial *Volvox*. Remember that members of the Volvocine series are probably related by common ancestral forms and that these modern genera did not necessarily evolve from each other.

Procedure 23.5
Examine members of the Volvocine line of algae

Use table 23.5 to record your observations as you sequentially examine each of the following organisms with your microscope:

- *Chlamydomonas* is among the most primitive and widespread of the green algae. It is a unicellular biflagellate alga (fig. 23.2). Most species of *Chlamydomonas* are isogamous, which means that all of their gametes are identical in size and appearance. All species of the Volvocine line consist of cells similar to *Chlamydomonas*, but the cells are in different configurations.

- *Gonium* is the simplest colonial member of the Volvocine line (fig. 23.3). A *Gonium* colony consists of 4, 8, 16, or 32 *Chlamydomonas*-like cells held together in the shape of a disk by a gelatinous matrix. Each cell in the *Gonium* colony can divide to produce cells that produce new colonies. Like *Chlamydomonas*, *Gonium* is isogamous.

TABLE 23.5

EVOLUTIONARY SPECIALIZATION OF MEMBERS OF THE VOLVOCINE LINE

Characteristic	Type of Algae				
	Chlamydomonas	*Gonium*	*Pandorina*	*Eudorina*	*Volvox*
Number of cells					
Colony size					
Structural and functional specializations of cells					
Reproductive specialization (isogamy vs. oogamy)					

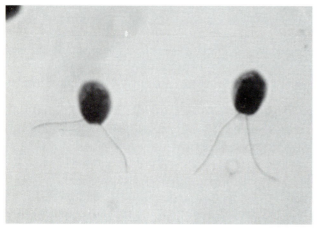

FIGURE 23.2

Chlamydomonas.

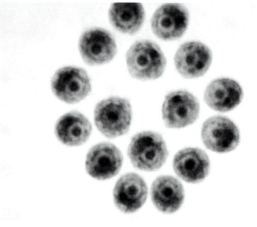

FIGURE 23.3

Gonium.

Question 6

a. Why do colonies of *Gonium* consist of only 4, 8, 16, or 32 cells? _____

_____

That is, why are there no 23-celled colonies? _____

_____

• *Pandorina* consists of 16 or 32 *Chlamydomonas*-like cells held together by a gelatinous matrix (fig. 23.4). Examine how *Pandorina* moves. Flagella on *Pandorina* move this ellipsoidal alga through the water like a ball. Eyespots are larger in cells at one end of the colony than in cells at the other end. After attaining its maximum size, each cell of the colony divides to form a new colony. The parent matrix then breaks open like Pandora's box (hence the name *Pandorina*) and releases the newly formed colonies. *Pandorina* is isogamous.

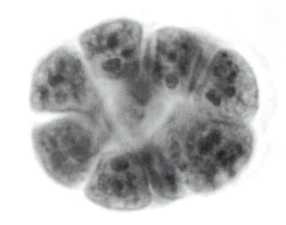

FIGURE 23.4

Pandorina.

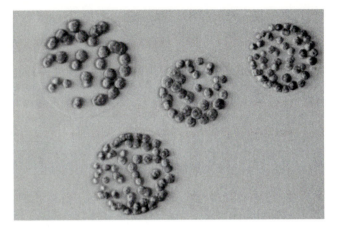

FIGURE 23.5

Eudorina.

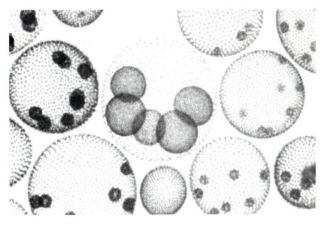

FIGURE 23.6

Volvox.

Question 7

a. What is the significance of a specialization at one

end of a colony? _____

_____

- *Eudorina* is a spherical colony comprised of 32, 64, or 128 cells (fig. 23.5). Cells in a colony of *Eudorina* differ in size; smaller cells are located at the anterior part of the colony. These smaller cells cannot reproduce to form new colonies. The anterior surface is determined by the direction of movement.

Question 8

a. What is the significance of the structural and

functional specializations of *Eudorina?* _____

_____

- *Volvox* is the largest and most spectacular organism of the Volvocine line. *Volvox* is a spherical colony made of thousands of vegetative cells and a few reproductive cells (fig. 23.6). *Volvox* is oogamous (i.e., the female gametes are large, nonmotile egg cells) and polar. Polar organisms have anterior and posterior poles (relative to their direction of movement). Flagella spin the colony on its axis. In some species of *Volvox* and *Gonium*, cytoplasmic strands form a conspicuous network among the cells.

Question 9

a. Does the *Volvox* colony spin clockwise or

counterclockwise? _____

EVOLUTION OF BODY FORM

Division Chlorophyta also provides several examples demonstrating the evolution of body form. Examine the algae on demonstration and take special note of their differing body forms:

- *Ulothrix* is a simple, unbranched, filamentous alga with cellular division in one plane.

- *Stigeoclonium* is an elongate, branching, filamentous alga with additional cellular divisions in a second plane, but in only a few cells.

- *Percursaria* is a broadened, flattened, filamentous alga that is one cell thick. It is formed when each cell divides in two planes.

- *Ulva* has a "leafy" plant body that results from cellular division in three planes. Because division in the third plane is restricted to a single division, the flattened blade is only two cells thick.

Question 10

a. How could evolutionary changes in planes of cellular

division produce different body forms? _____

_____

b. What would be the body form of an organism whose

cells divide randomly in one plane? _____

two planes? _____

three planes? _____

Summary of Activities

1. Review the terminology associated with Mendelian genetics.
2. Establish a parental population and calculate its allelic and genotypic frequencies.
3. Verify the Hardy-Weinberg equilibrium of genotypic frequencies in the offspring of the parental population.
4. Determine the effects of 100% negative selection pressure on allelic frequencies.
5. Determine the effects of 20% negative selection pressure on allelic frequencies.
6. Examine members of the Volvocine line of algae that have variations in colony complexity.
7. Examine genera of green algae with variations in body form.

Questions for Further Thought and Study

1. How would selection against heterozygous individuals affect the frequencies of homozygous individuals?

2. How are genetic characteristics associated with nonreproductive activities such as feeding affected by natural selection?
3. Why do you think *Volvox* is considered an evolutionary "dead end"?

Doing Botany Yourself

Design an experiment to determine the phylogenetic relationship among the Volvocine line of algae. What information about their DNA sequences would be useful?

Writing to Learn Botany

Hardy-Weinberg equilibrium assumes that pollination and subsequent fertilization must be random. Is this true for most wildflower populations? What characteristics of these plants influence pollination patterns?

Bacteria and Cyanobacteria

Objectives

By the end of this exercise you will be able to
1. describe distinguishing features of members of the kingdom Monera
2. describe differences between bacteria and cyanobacteria
3. identify representative examples of bacteria and cyanobacteria
4. perform a Gram stain.

Members of kingdom Monera are the most ancient organisms on earth and include major groups called bacteria and cyanobacteria (formerly known as blue-green algae). Monerans are structurally simpler than all other organisms because they lack membrane-bounded organelles such as nuclei, mitochondria, and chloroplasts. Cells lacking membrane-bounded organelles are **prokaryotic,** and monerans are **prokaryotes.** All other organisms are **eukaryotic** since their cells typically contain membrane-bounded organelles.

BACTERIA

Bacteria are distributed more widely than any other group of organisms. All individual bacteria are microscopic (1 mm or less in diameter); a single gram of soil may contain over a billion bacteria. Bacteria have cell walls that produce three characteristic shapes (fig. 24.1).

- Coccus (spherical)
- Bacillus (rod-shaped)
- Spirillum (spiral)

Most bacteria are **heterotrophic,** meaning that they derive their energy from organic molecules made by other organisms. Heterotrophic bacteria are **decomposers** because they feed on dead organic matter and release nutrients locked in dead tissue.

A laboratory culture of bacteria usually consists of a tube of liquid nutrients (broth) containing growing bacteria or a tube or plate of solidified agar with bacteria growing on the surface. The jellylike agar was melted, mixed with nutrients, and poured into tubes or plates to solidify. Many, but not all, species of bacteria are easily cultured in nutrient broth or on a layer of nutrient-rich agar.

Bacteria that derive their energy from photosynthesis or the oxidation of inorganic molecules are **autotrophic.** However, photosynthesis in bacteria is typically different from that of eukaryotes because sulfur rather than oxygen is produced as a by-product.

Question 1
a. What is the biological significance of decomposers?

_____

_____

b. What term best describes heterotrophic bacteria that feed on living tissue? _____

Bacteria reproduce asexually via **fission,** in which a cell's DNA replicates and the cell pinches in half without the nuclear and chromosomal events associated with mitosis. Some bacteria have genetic recombination via **conjugation,** in which all or part of the genetic material of one bacterium is transferred to another bacterium and a new set of genes is assembled.

Some bacteria can produce light when placed in favorable conditions in the presence of oxygen. Such

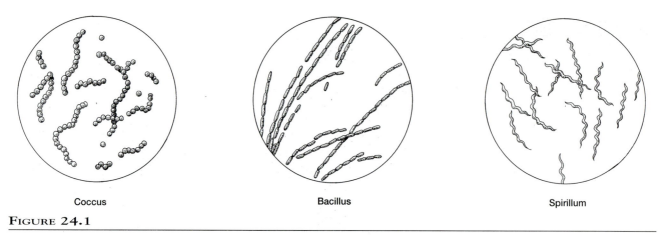

| Coccus | Bacillus | Spirillum |

FIGURE 24.1

Three shapes of bacteria.

bacteria are plentiful in seawater and often live symbiotically in light-organs of some deep-water fish.

Although some bacteria are pathogenic (causing diseases such as scarlet fever, pneumonia, syphilis, and tuberculosis, to name a few), most bacteria are harmless to humans. Indeed, many beneficial bacteria live in and on your body. Nevertheless, you should handle bacteria with care. The preparation of wet mounts of bacterial cultures requires proper use of a transfer loop and sterilizing flame. Your instructor will demonstrate this technique.

Procedure 24.1
Observe living cultures and prepared slides of bacteria

1. Obtain a slide, coverslip, transfer loop, alcohol burner, and a culture of living bacteria.

2. Available cultures should include the following bacteria, among others:

   Bacillus megaterium—a large bacterium that is resistant to radiation, desiccation, and heat

   Rhodospirillum rubrum—a photosynthetic purple bacterium

   Escherichia coli—found in the human intestine; the most intensively studied of all bacteria

   Staphylococcus epidermidis—a coccus found among the normal flora of skin

3. Using proper technique, apply a loop of bacteria to a drop of water on a slide. Cover with a coverslip.

4. Examine a wet mount of each of the bacterial species available.

5. Complete table 24.1 with observations of size and shape of the cells.

6. Examine any prepared slides of bacterial species that are available. Note the relative size and shape of the bacterial cells.

Procedure 24.2
Culture common bacteria

1. Obtain a sterile cotton swab and a closed petri plate containing sterile nutrient agar.

2. Open the packaged swab and drag the tip over a surface such as your teeth, face, or tabletop.

3. Open the petri plate and drag the exposed swab over the surface of the agar in the manner demonstrated by your instructor.

4. Close the lid and tape it shut. Label the plate with a wax pencil and place it in an incubator or in a warm area.

5. After 24 to 48 hours examine the agar for bacterial growth. Record your observations.

Gram Stain

One of the most important techniques to classify bacteria is the **Gram stain,** which is based on the different structural and chemical compositions of bacterial cell walls. Gram staining is important because it often correlates with the sensitivity of a bacterium to antibiotics. **Gram-positive** bacteria (e.g., *Streptococcus*) have a thick cell wall that retains a purple dye, while **Gram-negative** bacteria (e.g., *Escherichia coli*) have a much thinner cell wall that does not retain the dye.

During the Gram stain technique, crystal violet and iodine are applied to stain all of the bacteria purple. Then alcohol is used to decolorize or remove the stain from the surface of the Gram-negative cell walls that do not bind the stain. Finally, safranin is used to counterstain the Gram-negative cells with a red color contrasting to purple Gram-positive cells. In the following exercise, you will perform a Gram stain on some of your bacteria to see the difference between Gram-negative and Gram-positive cells.

TABLE 24.1

THE RELATIVE SIZE AND SHAPE OF SOME COMMON BACTERIA

Species	Relative Size	Shape

Procedure 24.3

Prepare and observe Gram-stained bacteria

1. Use the wide end of a toothpick to scrape your teeth near the gum line.

2. Thoroughly mix what's on the tip of the toothpick in a small drop of water on a microscope slide.

3. Allow this bacterial smear to dry.

4. Heat the slide gently by holding it with a clothespin and passing it over the top of a flame three to four times. If hot plates are available, hold the slide to the hot surface for 10 seconds. This heat will make the bacteria adhere to the slide.

5. When the slide has cooled, cover it with crystal violet for 1 minute.

> **C A U T I O N**
>
> Be careful not to inhale or spill crystal violet or any other biological stain on your skin.

6. Gently rinse the slide with water.

7. Cover the slide with iodine for 1 minute.

8. Drop 95% alcohol on the smear with an eyedropper until no purple shows in the alcohol coming off the slide.

9. Rinse the slide with water to remove the alcohol.

10. Cover the smear with safranin for 1 minute.

11. Gently rinse the slide with water. Air-dry the slide, or blot gently if necessary.

12. Observe the smear with your microscope using low power and then high power and/or the oil-immersion objective.

13. Repeat the Gram-staining procedure using *Bacillus megaterium* that you obtain from a culture tube. Your instructor will demonstrate sterile technique to open, sample, and close the culture of bacteria.

Question 2

a. Which type of bacteria is most prevalent in the sample from your teeth? _____

b. Is *Bacillus megaterium* Gram-positive or Gram-negative? _____

Nitrogen Fixation by Bacteria

Certain bacteria and cyanobacteria transform atmospheric nitrogen (N_2) into other nitrogenous compounds that plants can use as nutrients. This process is **nitrogen fixation.** All organisms need nitrogen as a component of their nucleic acids, proteins, and amino acids. However, chemical reactions capable of breaking the strong triple bond between atoms of atmospheric nitrogen are limited to certain bacteria and cyanobacteria. This process uses an enzyme called nitrogenase along with ATP, energized electrons, and water to convert N_2 to ammonia (NH_3). Ammonia can be absorbed by plants and used to make proteins and other macromolecules.

Rhizobium is a bacterium capable of nitrogen fixation, and it can grow intimately with roots of some plants called legumes (e.g., clover, alfalfa, and soybeans). Such associations between *Rhizobium* and host roots form **nodules** on the roots. The nitrogen-fixers living inside the nodules provide ammonia to the plant while the plant provides sugars and other nutrients to the bacteria.

Procedure 24.4

Observe root nodules

1. Observe the root systems on display, and note the nodules.

2. Examine a prepared slide of a cross section of a nodule.

Question 3

a. Where are the bacteria? _____

Are they between cells or inside cells? _____

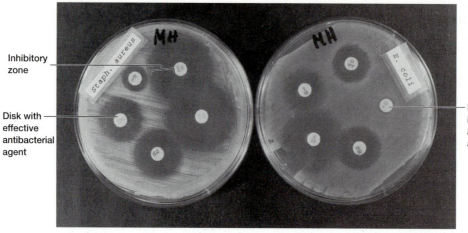

Inhibitory zone

Disk with effective antibacterial agent

Disk with ineffective antibacterial agent

FIGURE 24.2

Sensitivity plates. Bacterial growth is inhibited near the disks containing antibacterial agents. If the agents are ineffective, bacteria grows up to the disk.

b. Why is this relationship between a plant and

bacterium called a symbiosis? _____

_____

c. How does *Rhizobium* benefit from this association?

_____

_____

d. How does the host plant benefit from the association?

_____

_____

Economically Important Bacteria

Examine the living cultures of economically important bacteria in the lab. Most of these cultures are growing on a solid agar medium. As one or more microscopic bacterial cells reproduce on the surface of a solid medium, the new cells usually remain in contact with each other. Eventually the growing mass of cells becomes large enough to see easily; this group of cells is known as a **bacterial colony.** Bacterial colonies often have a characteristic shape, texture, or color. As you examine the demonstration cultures, take special note of their color, the appearance of the margin of the colony, and whether the bacterium grows on or in the growth medium. These characteristics are commonly used to identify bacteria.

Question 4
a. How do the colonies of bacterial cultures on display differ in appearance? (Consider texture, color,

density, colony shape, etc.) _____

_____

Bacterial Sensitivity to Inhibitors

Some bacterial species are more sensitive to growth inhibitors such as antibiotics than other species. For example, an antibiotic may inhibit growth of *Streptococcus* more than growth of *Staphylococcus*. This is important information to a physician who must select one of many available antibiotics to treat a bacterial infection.

To determine the most effective antibiotic, a medical laboratory technologist may set up a sensitivity plate. A **sensitivity plate** is a petri plate of solid medium uniformly inoculated on its entire surface with a known bacterium or an unknown sample from an infected patient. After inoculation, four to eight small paper disks are placed on the agar surface. Each disk was soaked in a different antibiotic. After 24 hours, an effective antibiotic will produce a visible halo of clear surface around the disks where it inhibited growth of the bacteria (fig. 24.2). If the antibiotic was ineffective, the bacteria grows to the edge of the paper disk.

You've probably seen television commercials for products such as mouthwash or disinfectant that "kills germs on contact." Many of these products are developed and tested using sensitivity plates. The mouthwash is effective if no bacteria grows around a paper disk that was soaked in the mouthwash.

Examine the sensitivity plates of bacterial cultures on demonstration.

Procedure 24.5
Examine sensitivity plates

1. Obtain from your instructor one of each type of sensitivity plate that has been set up.
2. Examine each plate, and note the bacteria used to inoculate the plate and the kinds of disks distributed on the plate.

24-4

TABLE 24.2

INHIBITION OF FOUR BACTERIAL SPECIES BY VARIOUS GROWTH INHIBITORS

Antibiotic	Plate 1 _____	Plate 2 _____	Plate 3 _____	Plate 4 _____

3. Determine which disks inhibited the bacteria strongly, weakly, or not at all.

4. Record the bacterial species and your observations in table 24.2.

Question 5

a. Based on their appearance, which drugs or chemicals retard the growth of bacteria? _____

_____

CYANOBACTERIA: BLUE-GREEN ALGAE

Cyanobacteria, like bacteria, are prokaryotes that grow in many environments. Cyanobacteria are photosynthetic, and their pigments include **chlorophyll *a*** and accessory pigments **phycocyanin** (blue) and **phycoerythrin** (red). Because these pigments vary in proportion, only about half of the cyanobacteria are actually blue-green in color; many cyanobacteria are brown to olive green.

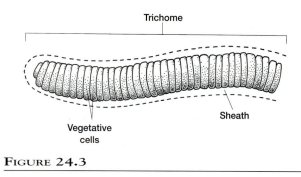

FIGURE 24.3

Oscillatoria.

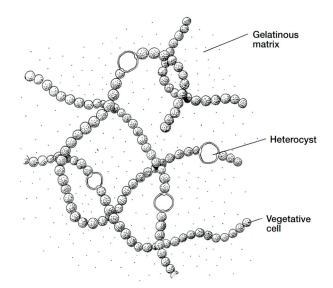

FIGURE 24.4

Nostoc.

Cyanobacteria reproduce by fission and are often surrounded by a jellylike **sheath.** Since cyanobacteria are prokaryotes, they are not related to other algae, which are all eukaryotic. Cyanobacteria such as *Oscillatoria* are often responsible for many of the disagreeable tastes, colors, and odors in water.

Procedure 24.6
Examine cyanobacteria

1. Examine living material and prepared slides of *Oscillatoria* with your microscope (fig. 24.3). *Oscillatoria* grows as trichomes, which are long chains of cells.

Question 6

a. Do all cells of a trichome of *Oscillatoria* appear

similar? _____

Explain. _____

_____

2. Examine living material and a prepared slide of Nostoc, commonly called witch's butter or starjelly (fig. 24.4). *Nostoc* forms large, grapelike colonies. Trichomes of *Nostoc* consist of small vegetative cells and larger, thick-walled **heterocysts,** in which nitrogen fixation occurs.
3. Examine a wet mount of living *Gloeocapsa*, which has a thick, gelatinous sheath (fig. 24.5).
4. Add a drop of dilute India ink to the slide of *Gloeocapsa* so that the sheath will stand out against the dark background. Often *Gloeocapsa* forms clusters of cells and therefore has a colonial body form. Locate one of these colonies.

Question 7

a. Do adjacent *Gloeocapsa* cells share a common

sheath? _____

b. What do you suppose is the function of the sheath?

_____

_____

c. Do clusters of *Gloeocapsa* represent multicellular

organisms? _____

Why or why not? _____

_____

_____

5. Examine some living *Merismopedia*, which also forms colonies. Sketch these cyanobacteria in the following space.

Question 8

a. How is the shape of *Merismopedia* different from other cyanobacteria you've studied in this exercise?

_____

_____

b. How would a colony attain this shape? _____

_____

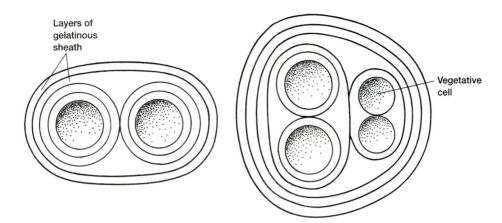

Layers of gelatinous sheath

Vegetative cell

FIGURE 24.5

Gloeocapsa.

Summary of Activities

1. Observe living cultures and prepared slides of *Bacillus megaterium*, *Rhodospirillum rubrum*, and *Escherichia coli*.
2. Prepare and observe a Gram stain.
3. Observe the display of root systems with nodules.
4. Examine a prepared slide of a cross section of a nodule.
5. Examine the display of economically important bacteria.
6. Examine the display of bacteria sensitivity plates.
7. Examine living material and prepared slides of *Oscillatoria*, *Nostoc*, *Gloeocapsa*, and *Merismopedia*.

Questions for Further Thought and Study

1. What is meant by "Gram-positive"?
2. What happens when milk is pasteurized?
3. What causes milk to sour?

Doing Botany Yourself

Obtain two nutrient agar plates. Touch and drag the tip of your finger on the agar surface of one plate. Wash your hand and repeat the procedure on the other plate. Incubate the plates for 24 to 48 hours. Then compare colony appearances and number. What can you conclude about the presence and diversity of bacteria on your plates? Is colony appearance a good way to distinguish species of bacteria?

Writing to Learn Botany

There is a great diversity of roles in a typical ecosystem, some of which are shared by a variety of organisms. What ecological roles are performed by cyanobacteria?

Fungi

Objectives

By the end of this exercise you will be able to
1. describe the characteristic features of the kingdom Fungi
2. discuss variation in structures and sequence of events for sexual and asexual reproduction for three major divisions of the kingdom Fungi.

If asked to draw a fungus, most of us would find it difficult. We would probably draw a fuzzy, amorphic organism like a mold or a spongy, plantlike mushroom. Although fungi and plants share some characteristics such as cell walls and attachment to a substrate, fungi are unique and structurally diverse organisms that warrant classification as a separate kingdom.

The basic structure of a fungus is the **hypha** (pl., hyphae), a slender filament of cytoplasm and haploid nuclei enclosed by a cell wall (fig. 25.1). Hyphae of some divisions and species have crosswalls called **septa** that separate cytoplasm and one or more nuclei into cells.

Other hyphae have incomplete or no septa (i.e., are aseptate) and therefore are **coenocytic** (multinucleate). **Mycelium** is a collective term for a cottonlike mass of hyphae constituting an individual organism; it may extensively permeate soil, water, or living tissue. Notably, the cell walls of fungi are usually not cellulose but are made of **chitin,** the same polysaccharide in the exoskeleton of insects and crustaceans.

Fungi are absorptive **heterotrophs.** Through their chitinous cell walls fungi secrete enzymes for **extracellular digestion** of organics in the substrate, after which the digested nutrients are absorbed into the mycelium. Heterotrophs gain their energy from organic molecules made by other organisms. Most fungi gain nourishment from dead organic matter and are called **saprophytes.** Other fungi feed on living organisms and are **parasites.** Many parasitic fungi have modified hyphae called **haustoria,** which are thin extensions of the hyphae that penetrate into living cells and absorb nutrients.

ASEXUAL REPRODUCTION IN FUNGI

Fungi commonly reproduce asexually by mitotic production of haploid vegetative cells called spores produced in sporangia, conidiophores, and other related structures. Spores are microscopic and surrounded by a covering well suited for the rigors of distribution into the environment. *Pilobolus*, an interesting fungus, points its sporangia toward the sun. This orientation of an organism to light is called phototaxis. *Pilobolus* ejects its entire sporangium as far as 2 meters to distribute its spores.

Budding and fragmentation are two other methods of asexual reproduction. Budding is mitosis with an uneven distribution of cytoplasm and is common in yeasts. After budding, the cell with the lesser amount of cytoplasm eventually detaches and matures into a new organism. Fragmentation is the breaking of an organism into one or more pieces that develop into new individuals.

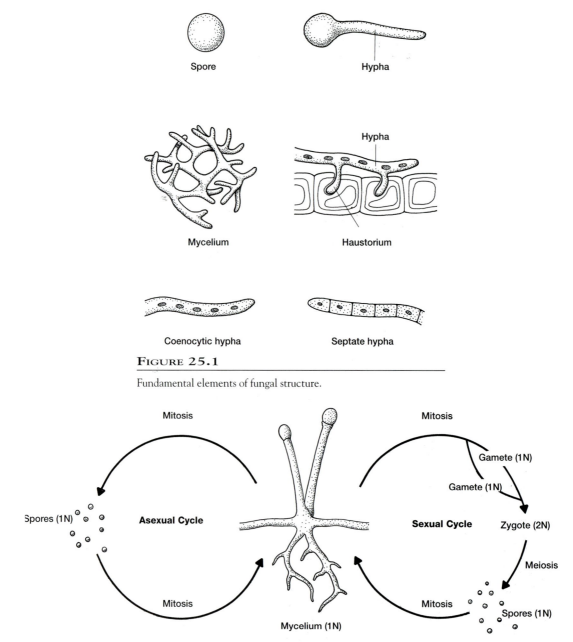

FIGURE 25.1

Fundamental elements of fungal structure.

FIGURE 25.2

Generalized life cycle of a fungus.

SEXUAL REPRODUCTION IN FUNGI

The sexual life history of fungi includes the familiar events of vegetative growth, genetic recombination, meiosis, and fertilization. However, the timing of these events is unique in fungi. Fungi reproduce sexually when hyphae of two genetically different individuals of the same species encounter each other. A generalized life cycle of fungi (fig. 25.2) illustrates four important features:

1. Nuclei of a fungal mycelium are haploid during most of the life cycle.

2. Gametes are produced by mitosis and differentiation of haploid cells rather than directly from meiosis of diploid cells.

3. Meiosis quickly follows formation of the zygote, the only diploid stage.

4. Haploid cells produced by meiosis are not gametes; rather, they are spores that grow into mature haploid organisms. Recall that asexual reproduction produces spores by mitosis. In both cases, haploid spores grow into mature mycelia.

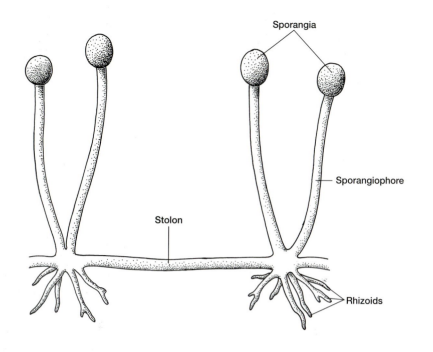

FIGURE 25.3

Vegetative and asexual reproductive structures of *Rhizopus*.

CLASSIFICATION OF FUNGI

The great diversity of fungi results from modifications of hyphae into varied and specialized reproductive structures often unique to a division, genus, or species. The three major divisions of fungi include Zygomycota, Ascomycota, and Basidiomycota. A fourth division, Deuteromycota, is an artificial rather than phylogenetic classification that includes species with no known sexual, or "perfect," phase. For this reason they are sometimes called Imperfect Fungi. As you examine members of these major groups, carefully note variations in the fundamental structure of vegetative mycelia and specialized structures associated with sexual and asexual reproduction.

Division Zygomycota (Bread Molds)

Zygomycetes (600 species), which include common bread molds, derive their name from resting sexual spores called zygospores that characterize the group. Most zygomycetes are saprophytic and their vegetative hyphae lack septa (i.e., they are aseptate).

Obtain a container of moldy bread that was moistened and exposed to air for a few days. Note the velvet texture and various colors of the molds. Use a stereomicroscope to examine the mycelia; notice that they grow as a tangled mass of hyphae.

A common genus of bread mold is *Rhizopus*. Its hyphae are modified into rhizoids (holdfasts), stolons (connecting hyphae), and sporangiophores (asexual reproductive structures), as shown in figure 25.3. Sporangiophores are upright

hyphal filaments supporting asexually reproducing sporangia. Within a sporangium haploid nuclei are separated by cell walls and become spores. These spores are released to the environment when the sporangium matures and breaks open.

Question 1

a. How many species of mold are on the bread? _____

b. Do any of the molds on the bread have hyphae modified as sporangiophores and sporangia? _____

c. Is pigment distributed uniformly in each mycelium?

_____

Where is the pigment concentrated in each mold?

_____

d. What is the adaptive significance of spores forming on ends of upright filaments rather than closer to the protective substrate? _____

_____

Obtain a pure culture of *Rhizopus* from your instructor. This culture is growing in a sealed petri plate containing nutrient-fortified agar. Do not remove the top of the plate because you may contaminate the culture and unnecessarily release spores into the room. Note that *Rhizopus* is dark; this is why the common name of *Rhizopus* is black bread mold. Examine the mold closely. Your instructor has prepared a demonstration slide of *Rhizopus* stained with lactophenol cotton blue for you to examine.

Your instructor has also prepared some cultures of *Rhizopus* (or *Mucor* or *Phycomyces*) for observing the structures

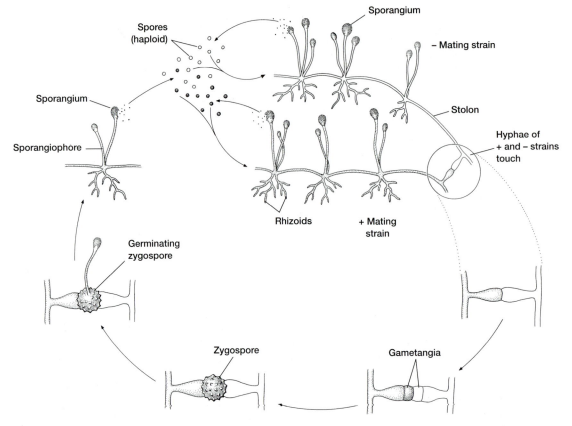

FIGURE 25.4

Life cycle of *Rhizopus*. Hyphae grow over the surface of the bread or other material on which the fungus feeds, producing erect, sporangium-bearing stalks in clumps. If both + and – strains are present in a colony, they may grow together, and their nuclei may fuse, producing a zygote. This zygote, which is the only diploid cell of the life cycle, acquires a thick, black coat called a zygosporangium (zygospore). Meiosis occurs during its germination, and vegetative, haploid hyphae grow from the resulting haploid cells.

formed during sexual reproduction. Obtain one of these plates. Because *Rhizopus* is isogamous (meaning that gametes from both strains look alike), the strains are not called male and female but are called + and –. On one side of the plate a + strain of *Rhizopus* was introduced, and on the other side a – strain was introduced. These strains grew toward each other and formed reproductive structures where they touched.

Sexual Reproduction in *Rhizopus*

1. Sexual reproduction begins when hyphae of each strain touch each other (fig. 25.4).

2. Where the hyphae touch, gametangia form and appear as swellings. Nuclei representing gametes from each strain differentiate within gametangia.

3. The walls separating the gametangia break down, and the nuclei fuse to form a zygote. Fusion of nuclei during sexual reproduction is sometimes referred to as conjugation; hence zygomycetes are called the "conjugating fungi."

4. A typically massive and elaborate zygospore differentiates around the zygote. Except for the zygote, all other nuclei are haploid.

5. Soon after the zygospore forms, the zygote undergoes meiosis. One or more of the resulting haploid cells soon germinates.

6. The hypha of the germinating cell breaks out of the zygospore, produces a sporangiophore, and forms spores asexually.

7. The spores are released and produce a new generation of mycelia.

Examine a prepared slide of *Rhizopus* with developed zygospores (fig. 25.5). Then observe the zygospores where the strains have touched. Be careful not to confuse zygospores with dark sporangia.

Question 2

a. In what structure is the dark pigment of *Rhizopus* concentrated? _____

b. Is *Rhizopus* reproducing sexually as well as asexually in the same petri plate? _____

How can you tell? _____

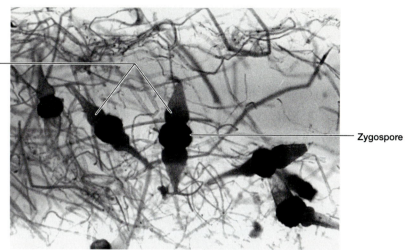

Gametangia

Zygospore

FIGURE 25.5

Sexual conjugation in *Rhizopus*.

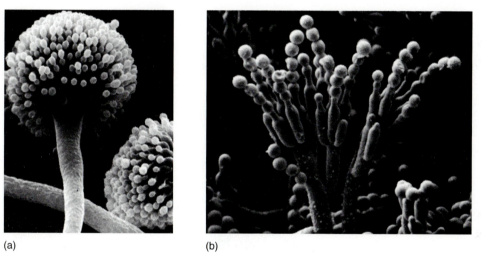

(a)   (b)

FIGURE 25.6

Conidiophores. (*a*) Characteristic conidiophores of *Aspergillus*, as viewed with the scanning electron microscope. (*b*) A colony of *Penicillium*, another important genus of Fungi Imperfecti, growing on an orange.

Division Ascomycota (Sac Fungi)

Division Ascomycota (30,000 species) includes yeasts, some molds, morels, and truffles. Its name is derived from a microscopic, sac-shaped reproductive structure called an ascus.

Ascomycetes reproduce asexually by forming spores called conidia. Modified hyphae called conidiophores partition nuclei in longitudinal chains of beadlike conidia (fig. 25.6). Each conidium contains one or more nuclei. Conidia form on the surface of condiophores in contrast to spores that form within sporangia in *Rhizopus*. When mature, conidia are released in large numbers and germinate to produce new organisms. Observe a living culture of Aspergillus and note the rounded tufts of conidia.

Yeast is a common ascomycete that exhibits budding, another form of asexual reproduction. The yeast used to produce wine and beer is usually a strain of *Saccharomyces*

cerevisiae. The fungi growing naturally on grapes used for making wine may impart a unique flavor to a wine, more than does the yeast or specific variety of grapes.

Examine a stock culture of *Saccharomyces* (fig. 25.7). Then prepare a wet mount of yeast from a culture dispensed by your instructor. Only a small amount of yeast is needed to make a good slide.

Question 3

a.   Do you see chains of yeast cells produced by budding?

_____

b.   How is the structure of yeast hyphae different from

that of molds? _____

_____

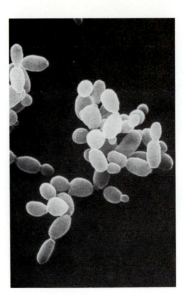

FIGURE 25.7

Yeast.

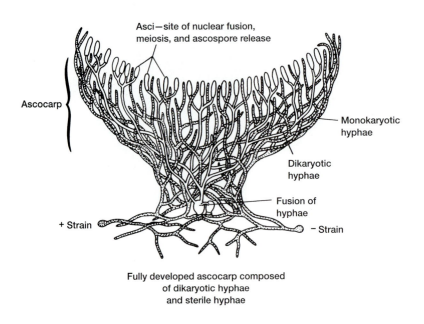

Asci—site of nuclear fusion, meiosis, and ascospore release

Ascocarp

Monokaryotic hyphae

Dikaryotic hyphae

Fusion of hyphae

+ Strain

− Strain

Fully developed ascocarp composed of dikaryotic hyphae and sterile hyphae

FIGURE 25.8

Sexual reproduction and ascocarp formation in ascomycetes.

Sexual Reproduction in Ascomycota

1. Sexual reproduction begins with contact of monokaryotic hyphae from two mating strains (fig. 25.8).

2. Where the hyphae touch, large multinucleate swellings appear and eventually fuse. Haploid nuclei of the two strains intermingle in the swelling.

3. A dikaryotic mycelium grows from this swelling. Each dikaryotic cell has one nucleus from each parent; the nuclei do not fuse immediately.

4. Tightly bundled dikaryotic hyphae grow and mingle with monokaryotic hyphae from each parent to form a cup-shaped ascocarp.

5. Dikaryotic cells lining the inside of the ascocarp form sac-shaped asci (sing., ascus).

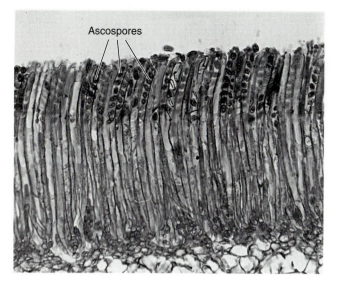

Ascospores

FIGURE 25.9

Asci from the lining of the cup of *Peziza*, a cup fungus. The diploid (2N) zygote in each ascus divides meiotically to produce four haploid (N) nuclei. Each of these nuclei then divides mitotically to produce a column of eight ascospores in each ascus.

6. The nuclei fuse in each ascus to form a zygote.

7. After fusion, meiosis produces four haploid ascospores.

8. Subsequent mitosis produces eight ascospores within each mature ascus.

9. The asci on the surface of the ascocarp rupture and release ascospores into the environment.

10. Each ascospore can produce a new mycelium.

Examine a prepared slide of a cross section through the ascocarp of *Peziza* (fig. 25.9).

Question 4

a. What is the difference between dikaryotic and

diploid cells? _____

_____

Division Basidiomycota (Club Fungi)

Basidiomycetes (25,000 species) are probably the most familiar fungi. They include mushrooms, puffballs, shelf fungi, and economically important plant pathogens such as rusts and smuts (see plate VIII, fig. 14). *Agaricus campestris* is a common field mushroom, and its close relative *A. bisporus* is cultivated for 60,000 tons of food per year in the United States. However, just one bite of *Amanita phalloides*, the "destroying angel," may be fatal.

Mushrooms are familiar examples of the above-ground portions of extensive mycelia permeating the soil. Examine a mushroom and other representative basidiomycetes on display and note the mushroom's cap, pileus, and gills on the undersurface of the cap (fig. 25.10). Gills are lined with microscopic, club-shaped cells called basidia, where sexual reproduction occurs. Division Basidiomycota is sometimes called the "club fungi" and derives its name from these characteristic basidia.

Sexual Reproduction in Basidiomycota

1. Haploid hyphae from different mating strains permeate the substrate (fig. 25.10; see also plate VIII, fig. 14).

2. Septa form between the nuclei in the hyphae and form monokaryotic primary mycelia. A monokaryotic mycelium has one nucleus in each cell.

3. Cells of the primary mycelia of different mating strains touch, fuse, and produce a dikaryotic secondary mycelium.

4. The secondary mycelium grows in the substrate.

5. The dikaryotic hyphae of the secondary mycelium eventually coalesce and protrude above the substrate as a tight bundle of hyphae called a basidiocarp (mushroom).

6. The basidiocarp forms a pileus, cap, and gills.

7. Dikaryotic, club-shaped basidia form on the surface of the gills.

8. The two nuclei in each basidium fuse to form a diploid zygote.

9. The zygote soon undergoes meiosis and produces haploid basidiospores.

10. The basidiospores are released from the basidia lining the gills and are dispersed by wind.

11. A basidiospore germinates into a new mycelium.

Examine a prepared slide of gills from the cap of *Coprinus*, a common mushroom (fig. 25.11). Note the dark basidiospores in rows along the surface of the gills. Interestingly, the gills form perpendicular to the ground and allow spores to free-fall easily and disperse. Recent research suggests that gravity determines the orientation of gills.

Question 5

a. How many spores would you estimate are present on the gills of a single cap of *Coprinus?* (Remember that

a prepared slide shows only a cross section.) _____

_____

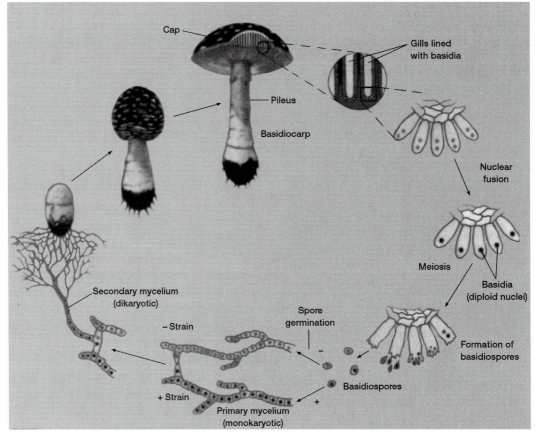

FIGURE 25.10

Life cycle of a basidiomycete. In primary mycelia, each cell has only one nucleus; in secondary mycelia, which are formed by the fusion of primary mycelia (plasmogamy), there are two nuclei, one derived from each of the strains that gave rise to the secondary mycelia, within each cell. Secondary mycelia ultimately may become interwoven, forming the basidiocarp, with gills. In these basidia meiosis immediately follows syngamy.

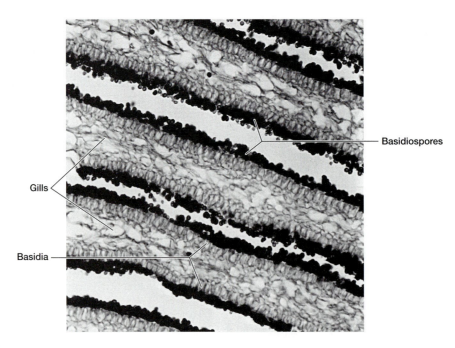

FIGURE 25.11

Gills of the mushroom *Coprinus*, a basidiomycete.

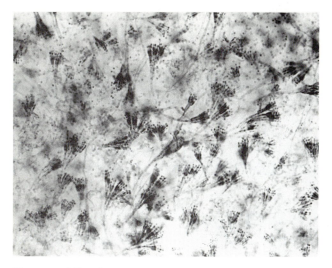

FIGURE 25.12

Penicillium with conidia.

Division Deuteromycota (Imperfect Fungi)

Deuteromycetes include 25,000 species that are difficult to classify because biologists have never induced them to reproduce sexually in the laboratory or seen sexual morphology in naturally growing specimens. These fungi have apparently lost their ability to reproduce sexually. Many species of *Penicillium* and *Aspergillus* are classified as Imperfect Fungi even though other members of these genera are considered ascomycetes. *Penicillium* reproduces asexually by producing conidia in a fashion similar to *Aspergillus*. Examine a prepared slide of *Penicillium* (fig. 25.12) and a pure, living culture provided by your instructor, and notice the formation of conidia.

Question 6

a. What is the relative size of *Penicillium* hyphae

compared with *Rhizopus* hyphae? _____

_____

Many deuteromycetes are economically important. For example, species of *Penicillium* are used to produce antibiotics. *Penicillium roquefortii*, which grows only in caves near Roquefort, France, gives a unique flavor to Roquefort cheese. *Aspergillus oryzae* is used to brew Japanese saki and to enrich food for livestock.

LICHENS

Lichens (25,000 species) are common, brightly colored organisms in which an ascomycete (rarely other fungi) live symbiotically with a photosynthetic alga or cyanobacterium. Symbiosis means living in a close and sometimes dependent association. About twenty-six genera of algae occur in different species of lichens. However, the green algae *Trebouxia* and *Trentepohlia* and the cyanobacterium *Nostoc* are in 90% of lichen species. Examine a prepared slide of a cross section of a lichen thallus, and note the intimate contact between the symbionts (fig. 25.13).

Lichens reproduce asexually by releasing fragments of tissue or specialized, stress-resistant packets of fungal and algal cells. Each of the two components (fungus and alga) may reproduce sexually by mechanisms characteristic of their division, and the new organisms may continue the lichen association.

The durable construction of fungi, linked with the photosynthetic properties of algae, enable lichens to proliferate into the harshest terrestrial habitats. Examine the dried lichens on display and note the three basic growth forms: crustose, foliose, and fruticose (fig. 25.14). The thallus of crustose lichens grows close to the surface of a hard substrate such as rock or bark. The lichen is flat and two-dimensional. Foliose lichens adhere to their substrate, but some of the thallus peels and folds away from the substrate in small sheets. Fruticose lichens are three-dimensional and often grow away from the substrate with erect stalks. The tips of the stalks are often sites of ascus formation by the sexually reproducing ascomycete symbiont.

Lichens are extremely sensitive to air pollution. This is probably because they are adapted to efficiently absorb nutrients and minerals from the air. This makes lichens particularly susceptible to airborne toxins.

Question 7

a. What is the value of photosynthetic algae to the

growth of a fungus in a lichen? _____

b. Would you expect lichens to grow best in rural or

urban environments? _____

Why? _____

_____

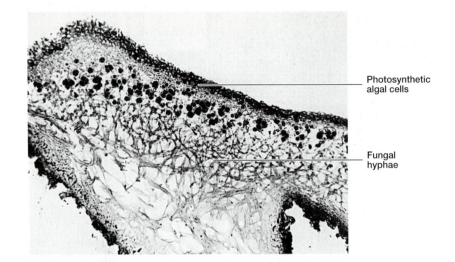

Photosynthetic
algal cells

Fungal
hyphae

FIGURE 25.13

Lichen thallus, cross section. Lichens consist of algae and fungi growing symbiotically.

(a)   Crustose lichen

(b)   Fruticose lichen   (c)   Foliose lichen

FIGURE 25.14

Growth forms of lichens: (*a*) Crustose. (*b*) Fruticose. (*c*) Foliose.

25-10

MAKING WINE

Alcoholic fermentation by yeast is easily demonstrated (see exercise 8). Many biologists as well as nonbiologists use fermentation to make wine. For an introduction to home wine making, try the following procedure:

1. Thoroughly clean and sterilize all glassware.

2. Combine a cake of yeast with either bottled grape juice or cranberry juice. Mix the yeast and juice in a ratio of approximately 5 liters of juice to 1 gram of yeast.

3. Add approximately 650 mL of the juice-yeast mix to each of four 1-liter Ehrlenmeyer flasks (or use 1- or 2-liter recycled plastic pop bottles).

4. Dissolve the following amounts of sucrose in each flask:

Flask 1: 75 g

Flask 2: 150 g

Flask 3: 300 g

Flask 4: no sucrose

5. Set up the fermentation apparatus as shown in figure 25.15 for each flask.

6. Be sure to keep the cultures anaerobic by keeping the end of the exit tube under water in the adjacent smaller flasks. This prevents contamination by airborne bacteria and yeast.

7. Incubate the flasks at temperatures between 15° and 22°C. Although fermentation will continue for a month or so, most fermentation occurs within the first 14 days. Fermentation is complete when bubbling stops.

8. To test your wine, remove the stopper and use a piece of tubing to siphon off the wine solution without disturbing the sediment in the bottom of the flasks. You may then want to filter the solution to remove any remaining yeast cells from the wine.

9. Taste your wine. What differences are there in wines produced with different amounts of sugar? If your wine has been contaminated by bacteria that produce acetic acid, vinegar may have been formed, so take your first sip slowly.

10. If you're interested in the finer points of wine making, visit your local bookstore or library. There may also be a local society of amateur wine makers in your area that will be glad to give you some pointers on creating a wine with a "delightful bouquet."

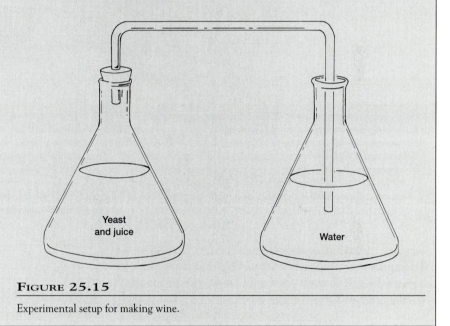

Yeast and juice

Water

FIGURE 25.15

Experimental setup for making wine.

Summary of Activities

1. Examine a variety of fungi on moldy bread.
2. Examine a live culture of asexually reproducing *Rhizopus*.
3. Examine a demonstration slide of *Rhizopus* stained with lactophenol cotton blue.
4. Examine a live culture of sexually reproducing *Rhizopus* with zygospores.
5. Examine a cross section of a *Peziza* ascocarp.
6. Examine living cultures of *Aspergillus*, *Penicillium*, and *Saccharomyces*.
7. Examine mushrooms on display.
8. Examine a prepared slide of *Coprinus* gills.
9. Examine lichens on display and a cross section of a lichen thallus.

Questions for Further Thought and Study

1. Mushrooms often sprout from soil in rows or circles called fairy rings. How would you explain this formation?
2. What advantages does asexual reproduction have over sexual reproduction?

3. Does dominance of the haploid condition in a fungal life cycle offer an adaptive advantage? Why or why not?

4. Compare and contrast the structure of a fungal mycelium with the structure of a filamentous alga.

5. What is the advantage of maintaining a dikaryotic condition rather than immediate nuclear fusion?

6. In fungi, the only distinction between a spore and a gamete is function. Explain this difference.

Doing Botany Yourself

Wine making is a multimillion dollar industry. The biology of wine making, however, is simple. Try making your own special kind of wine by following the instructions in the later part of this exercise. Good luck!

Writing to Learn Botany

Compare and contrast the fundamental life cycle of a fungus with that of plants and animals. When does meiosis occur in the sequence of events? Which stages are haploid and which are diploid?

Algae

Objectives

By the end of this exercise you will be able to

1. discuss the distinguishing features of different groups of algae
2. discuss the economic importance of algae
3. understand "alternation of generations" in green algae.

THE KINGDOMS OF EUKARYOTIC ORGANISMS

Eukaryotes are organisms comprised of cells having membrane-bounded nuclei. Kingdom Fungi (see exercise 25) was the first of three kingdoms that include plants and plantlike organisms (kingdoms Fungi, Plantae, and Protoctista). Fungi have cell walls and are heterotrophic; they feed on organic matter produced by other organisms because fungi cannot make their required organic compounds from inorganic substances.

Plants are multicellular, **autotrophic** organisms, meaning that they can synthesize all required organic compounds from inorganic substances using external energy, which is usually sunlight.

Protoctista is the oldest and most diverse kingdom of eukaryotes. In a sense, protoctists include all eukaryotes that lack the distinguishing characteristics of plants, animals, and fungi. Protoctists include simple eukaryotes, such as amoebas, and some unusual multicellular forms, such as the brown alga (kelp). Protoctists are mostly microscopic, unicellular organisms and probably share a common ancestry with fungi, multicellular plants, and animals. Members of kingdom Protoctista include three general groups: algae, protozoans, and slime molds. You will study algae and slime molds in this exercise.

THE ALGAE

Algae are photosynthetic, eukaryotic organisms typically lacking multicellular sex organs. The major groups of algae are distinguished in part by their energy storage products, cell walls, and color, resulting from the type and abundance of **pigments** (substances that absorb light) in their plastids (table 26.1). Thus, we refer to the major groups of algae as green algae, brown algae, golden brown algae, and red algae. Red algae owe their color to water-soluble pigments called **phycobilins.** Other algal pigments such as **chlorophylls** and **carotenoids** are insoluble in water but can be extracted with organic solvents such as acetone and alcohol.

Algae are also distinguished by their cellular organization. **Unicellular** algal species occur as single, unattached cells that may or may not be motile. **Filamentous** algal species occur as chains of cells attached end to end. These filaments may be few to many cells long and may be unbranched or branched in various patterns. **Colonial** algae occur as groups of cells attached to each other in a nonfilamentous manner. For example, a colony may include several to many cells adhering to each other as a sphere, flat sheet, or other three-dimensional shape. The cells may be embedded in a common matrix of gelatin secreted from the cells or attached by thin strands of cytoplasm. All the cells

TABLE 26.1

CHARACTERISTICS OF ALGAE BY DIVISION

Division	Predominant Cellular Organization	Common Pigments*	Storage Products	Cell Wall Components
Chlorophyta	Unicellular, filamentous, colonial	Chlorophyll *b*	Starch	Mainly cellulose
Phaeophyta	Filamentous, multicellular	Chlorophyll *c*, fucoxanthin	Laminarin, mannitol	Cellulose, alginates
Rhodophyta	Multicellular	Phycobilins	Modified starch	Cellulose, agar, carrageenan
Chrysophyta Class Bacillariophyceae	Unicellular	Chlorophyll *c*, fucoxanthin	Chrysolaminarin, lipids	Silica, calcium carbonate
Pyrrhophyta	Unicellular	Chlorophyll *c*	Starch, lipids	Pectin and cellulose plates
Euglenophyta	Unicellular	Chlorophyll *b*	Paramylon, lipids	Protein

*In addition to chlorophyll *a*.

of a colony are similar in structure and function and are metabolically independent. **Multicellular** organization is not typical of protists but describes algae of more complex design than simple colonies. Multicellular species have cells of different kinds and functions and show significant interdependence.

DIVISION CHLOROPHYTA (GREEN ALGAE)

Green algae are the most diverse and familiar algae in freshwater. However, a few genera live in saltwater. Although the name Chlorophyta means "green-chlorophyll" plants (chloro + phyta); green algae are not classified as plants. Green algae are considered to be ancestral to land plants and share many characteristics with land plants, such as

- chlorophyll *a*, which occurs in algae and green plants;
- chlorophyll *b*, which occurs in land plants and in the green and euglenoid algae;
- starch as the carbohydrate storage material;
- cell walls made of cellulose.

Let's examine a few representatives of green algae.

Unicellular Green Alga: *Chlamydomonas*

Chlamydomonas is a motile, unicellular alga found in soil, lakes, and ditches. It probably has the most primitive structure and type of reproduction among green algae. The egg-shaped cells of *Chlamydomonas* contain a large chloroplast and a pyrenoid, which produces and stores starch.

Procedure 26.1
Observe *Chlamydomonas*

1. Observe a drop of water containing living *Chlamydomonas*; note the movement of the cells.
2. If the movement is too fast, make a new preparation by placing 1 or 2 drops of methylcellulose on a slide and adding a drop of water containing *Chlamydomonas*.
3. Mix gently and add a coverslip.
4. Note the stigma, which appears as a reddish, light-absorbing spot at the anterior end of the cell.
5. Identify other structures shown in figure 26.1.
6. Examine prepared slides of *Chlamydomonas*.

Question 1
a. Is the movement of *Chlamydomonas* smooth or does it appear jerky? _____

b. Can you see both flagella? (You may need to reduce the light intensity to see flagella.) _____

c. How does methylcellulose affect movement of *Chlamydomonas*? _____
_____

d. How does the stigma help *Chlamydomonas* survive?
_____
_____

Asexual and Sexual Reproduction in *Chlamydomonas*

Chlamydomonas usually reproduces asexually via mitosis. Sexual reproduction in *Chlamydomonas* is a response to unfavorable environmental conditions. For sexual reproduction,

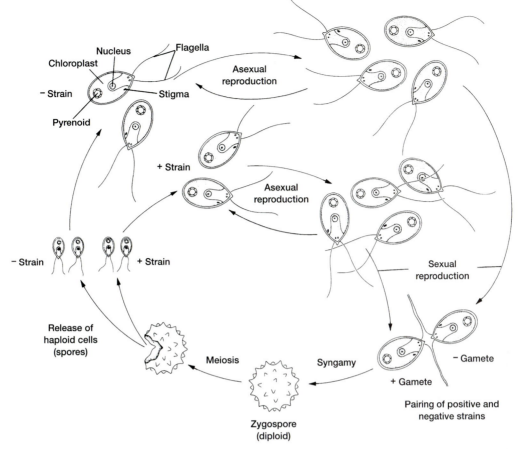

FIGURE 26.1

Chlamydomonas life cycle. Individual cells of *Chlamydomonas* (a microscopic, biflagellated alga) are haploid and divide asexually, producing copies of themselves. At times, such haploid cells act as gametes, fusing to produce a zygote, as shown in the lower right of the diagram. The zygote develops a thick, resistant wall, becoming a zygospore. Meiosis then produces four haploid individuals. Only + and − strains can mate with one another, although both may also divide asexually and reproduce themselves.

vegetative cells of *Chlamydomonas* undergo mitosis to produce gametes. The gametes fuse to form a **zygote,** which is the resting stage of the life cycle. In most species of *Chlamydomonas*, the male and female gametes are **isogamous,** meaning they have an identical shape and appearance. For convenience, isogametes of *Chlamydomonas* are referred to as + or −. Gametes unite to form a diploid zygote. **Syngamy** is the fusion of haploid gametes to form diploid cells. Under favorable conditions the zygote undergoes meiosis to produce spores. **Spores** are reproductive cells capable of developing into an adult without fusing with another cell. Study the life cycle of *Chlamydomonas* shown in figure 26.1.

Procedure 26.2
Observe syngamy

1. Place drops of + and − gametes of *Chlamydomonas* provided by your instructor next to each other on a microscope slide, being careful not to mix the two drops. Do not add a coverslip.

2. While you observe the drops through low power of the microscope, mix the two drops of isogametes.

3. Switch to high magnification and note the clumping gametes. Try to locate cells that have paired.

Question 2
a. Under what environmental conditions would a zygote not undergo meiosis immediately? _____

_____

b. Are spores of *Chlamydomonas* haploid or diploid?

_____

c. Which portions of the life cycle of *Chlamydomonas* are haploid? _____

d. Which are diploid? _____

_____

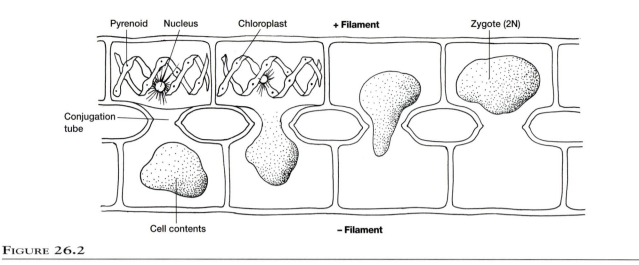

FIGURE 26.2

Conjugation in *Spirogyra*.

Filamentous Green Algae: *Spirogyra* and *Cladophora*

Spirogyra grows in running streams of cool freshwater and secretes a mucilage that makes it feel slippery. Observe living cultures and prepared slides of *Spirogyra*. Notice the shape of the chloroplasts and whether filaments are branched.

    Spirogyra has nonflagellated isogametes. During sexual reproduction (conjugation), filaments of opposite mating types lie side by side and form projections that grow toward each other. These projections touch and the separating wall dissolves, thus forming a **conjugation tube** (fig. 26.2). The cellular contents of the – strain then migrate through the conjugation tube and fuse with that of the nonmotile + strain. The resulting zygote develops a thick, resistant cell wall and is termed a zygospore. The **zygospore** is released when the filament disintegrates and then undergoes meiosis to form haploid cells that become new filaments.

    Reexamine the living culture of *Spirogyra* with your microscope. In the space below, sketch a filament of *Spirogyra*. Note the shape of its chloroplasts.

Question 3

a.   Are the filaments of *Spirogyra* branched? _____

_____

b.   What is the shape of the chloroplasts of *Spirogyra*?

_____

c.   Can you see any conjugation tubes? (If you can't, examine the prepared slides that show these

structures.) _____

d.   How do you think that *Spirogyra* reproduces

asexually? _____

_____

    Examine some living cultures and prepared slides of *Cladophora*. In the space below, draw a few cells showing their general shape and the filament's branching pattern.

Question 4

a.   How is *Cladophora* morphologically similar to

Spirogyra? _____

_____

How is it different? _____

_____

b.   What is the shape of its chloroplasts? _____

_____

    Mature *Cladophora* exists in diploid and haploid forms. The diploid stage of the life cycle produces spores and is called the **sporophyte.** The haploid stage of the life cycle produces gametes and is called the **gametophyte.** This phenomenon of alternating haploid and diploid life stages is called **alternation of generations,**

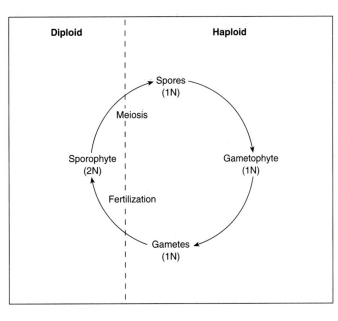

FIGURE 26.3

Alternation of generations.

and it is a reproductive cycle in which the haploid gametophyte produces gametes that fuse to form a zygote that germinates to produce a diploid sporophyte. Within the sporophyte, meiosis produces spores that germinate into gametophytes, thus completing the cycle (fig. 26.3). Alternation of generations occurs in many green algae, including *Cladophora*, and all land plants.

You should become familiar with the concept of alternation of generations because it occurs frequently in the plant kingdom and we will refer to it often. Refer to your textbook for more information.

Colonial Green Alga: *Volvox*

Volvox consists of many *Chlamydomonas*-like cells bound in a common spherical matrix. Each cell in the sphere has two flagella extending outward from the surface of the colony. Synchronized beating of the flagella spin the colony through the water like a globe on its axis. *Volvox* is one of the most structurally advanced colonial forms of algae, so much so that some biologists consider *Volvox* to be multicellular. Some of the cells of a *Volvox* colony are functionally differentiated; a few specialized cells can produce new colonies, and eggs and sperm are formed by different cells in the colony.

Volvox reproduces by **oogamy.** Motile sperm swim to and fuse with the large nonmotile eggs to form a diploid zygote. The zygote enlarges and develops into a thick-walled zygospore, which is released when the parent colony disintegrates. The zygospore then undergoes meiosis to produce haploid cells, which subsequently undergo mitosis and become a new colony. During asexual

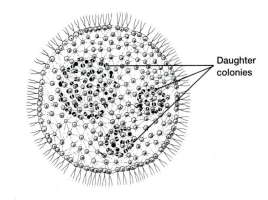

FIGURE 26.4

Volvox with interior daughter colonies.

reproduction, some cells of *Volvox* divide, bulge inward, and produce new colonies called **daughter colonies** that initially are held within the parent colony.

Procedure 26.3
Observe Volvox

1. Place a drop of living *Volvox* on a depression slide.

2. Using low magnification, observe the large, hollow, spherical colonies for a few minutes to appreciate their elegance and beauty (fig. 26.4).

3. Search for flagella on the surface.

4. Complement your observations of this alga by examining prepared slides of *Volvox*.

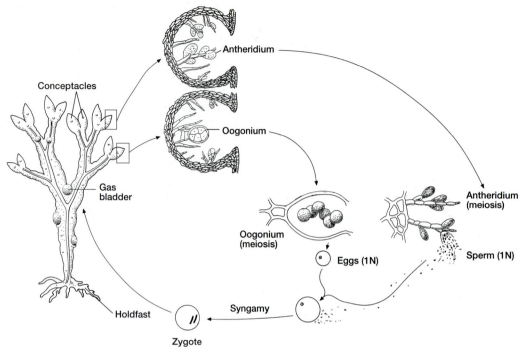

FIGURE 26.5

Life cycle of *Fucus*.

Question 5

a. What does oogamy mean? _____

_____

b. What are the tiny spheres inside the larger sphere of

Volvox? _____

_____

c. How do you suppose they get out? _____

_____

d. How do you think the number of cells in a young
Volvox colony compares with the number of cells in a

mature colony? _____

_____

DIVISION PHAEOPHYTA (BROWN ALGAE)

Phaeophytes are primarily marine algae that are structurally complex; there are no unicellular or colonial brown
algae. Brown algae usually grow in cool water and get their
name from the abundance of a brown pigment called **fu-
coxanthin.** Brown algae range in size from microscopic
forms to kelps over 50 m long. Thalli of some phaeophytes
are similar to those of land plants. Review table 26.1 for
the characteristics of brown algae.

Fucus

Fucus (rockweed) typically attaches to rocks in the inter-
tidal zone via a specialized structure called a **holdfast.** The
outer surface of *Fucus* is covered by a gelatinous sheath.
Tips of *Fucus* branches, called **conceptacles,** may be swollen
and contain reproductive structures called oogonia (female)
and antheridia (male), as shown in figure 26.5. **Oogonia** are
multicellular sex organs that produce one or more eggs. **Antheridia** are multicellular sex organs that produce sperm.
Most protists do not have multicellular reproductive organs.

The life cycle of *Fucus* is similar to that of animals
and higher plants. The mature thallus is diploid, and cells
within reproductive structures undergo meiosis to produce
gametes, thereby skipping the multicellular haploid stage
common to many protists, plants, and fungi.

Procedure 26.4
Examine *Fucus*

1. Examine *Fucus* in the lab, referring to figure 26.5.
   Use your dissecting microscope to examine a cross
   section of the flattened, dichotomously branched
   thallus of *Fucus*.

2. Notice the presence of swollen areas on the thallus of
   Fucus.

3. Working in small groups, dissect one of these
   structures.

4. Examine prepared slides of antheridia and oogonia of
   Fucus.

Question 6

a. How does the structure of *Fucus* differ from the green algae that you examined earlier in this lab? _____

_____

b. Are all portions of the thallus photosynthetic? _____ How can you tell? _____

_____

c. Considering where *Fucus* lives, what do you think is the function of its gelatinous sheath? _____

_____

d. Are the swollen structures solid masses or are they empty? _____

Question 7

a. Are the gametes of *Fucus* isogamous or oogamous?

_____

b. How does the structure of tissue surrounding the reproductive structures compare with that of green algae? _____

_____

Economic Importance of Phaeophyta

Brown algae are of considerable economic importance. In the Orient some brown algae are used as food. One of these algae is *Laminaria*, a kelp marketed as "kombu." If available, taste a small piece of kombu. How would you describe its taste and texture?

Brown algae are also important sources of alginic acid, which is a hydrophilic substance (i.e., it absorbs large quantities of water). Alginic acid is used as an emulsifier in dripless paint, ice cream, pudding mixes, and cosmetics. Observe the products on display and, in the case of foods, read their contents labels.

DIVISION RHODOPHYTA (RED ALGAE)

Red algae get their color from the presence of red phycobilins in their plastids. Red algae typically live in warm marine waters. The thallus of a red alga can be attached or free-floating, filamentous, or parenchymatous (fleshy).

Polysiphonia and *Porphyra*

Examine living *Polysiphonia*, which are highly branched and filamentous. Gametophytes of these organisms are unisexual; that is, they are either male or female. Compare the structure of *Polysiphonia* with that of *Porphyra*. "Blades" of *Porphyra* consist of two layers of cells separated by colloidal material.

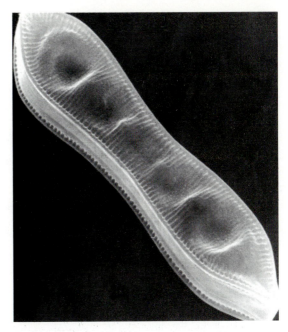

FIGURE 26.6

An electron micrograph of a diatom showing its ornate, silicon cell wall.

Economic Importance of Red Algae

Study the display of carrageenin, agar, and other products derived from red algae. Agar is a gelatinous polysaccharide used as a solidifying agent in culture media for microbiology labs. A 1% suspension of agar in heated water remains liquid until it cools to about body temperature. Pick up and feel a piece of agar, noting its texture.

DIVISION CHRYSOPHYTA CLASS BACILLARIOPHYCEAE (DIATOMS)

Diatoms are unicellular algae containing chlorophylls *a* and *c* and **xanthophyll** pigments that give them their golden brown color. Although diatoms are tiny, their great numbers, rapid rates of reproduction, and photosynthetic capacity make them a vital link in the food chain of the oceans.

Diatoms have hard cell walls made of silicon dioxide (glass) (fig. 26.6). These walls are arranged in overlapping halves, much like a petri plate. Examine a prepared slide of diatoms and a few drops of living diatoms. Sketch some of the cells. Some cells are long and thin whereas others are disklike.

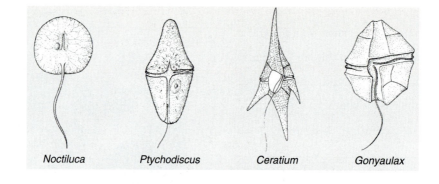

FIGURE 26.7

The dinoflagellates *Noctiluca*, *Ptychodiscus*, *Ceratium*, and *Gonyaulax*. *Noctiluca*, which lacks the heavy cellulosic armor characteristic of most dinoflagellates, is one of the bioluminescent organisms that cause the waves to sparkle in warm seas at certain times of year. In the other three genera, the shorter, encircling flagellum may be seen in its groove, with the longer flagellum projecting away from the body of the dinoflagellate.

Question 8

a. Can you see any pores in the walls of diatoms? _____

b. Are any of the diatoms moving? _____

c. If diatoms lack flagella, how do you explain their

motility? _____

_____

The glass walls of diatoms persist long after the remainder of the cell disintegrates. These walls may accumulate in layers of **diatomaceous earth** several hundred meters deep.

Procedure 26.5
Examine diatomaceous earth

1. Mount a small amount of diatomaceous earth in water on a microscope slide.

2. Examine it with your microscope.

3. Note the variety of shells, some broken and others intact. A mass of these shells is clean, insoluble, and porous.

Question 9

a. How would diatomaceous earth compare to sand as a

filter for a swimming pool? _____

Which would be better? _____

Why? _____

_____

DIVISION PYRRHOPHYTA (DINOFLAGELLATES)

Members of this division are all unicellular. Dinoflagellates are characterized by the bizarre appearance of their cellulose plates and by the presence of two flagella located in perpendicular grooves (fig. 26.7).

A "red tide" is caused by blooms of a red-pigmented dinoflagellate called *Ptychodiscus brevis*. The production of toxins and depletion of oxygen by these blooms of algae can kill massive numbers of fish. Dinoflagellates are important primary producers in oceans (second only to diatoms), and include many autotrophic and heterotrophic forms. Some dinoflagellates are bioluminescent, and others live symbiotically with corals.

Observe living cultures and prepared slides of *Peridinium* or *Ceratium*. Look for longitudinal and transverse flagella and flagellar grooves.

Question 10

a. How do the shapes of dinoflagellates compare with those of other unicellular algae that you have

observed in this lab? _____

_____

DIVISION EUGLENOPHYTA (EUGLENOIDS)

This small division includes mostly freshwater unicellular algae. Although plastids of euglenoids contain chlorophylls *a* and *b* (like the green algae), euglenoids are distinctive because their cell walls consist largely of protein. This protein makes the cell more flexible. Euglenoids are motile and have two flagella. Examine a prepared slide of *Euglena*.

Procedure 26.6
Observe living *Euglena*

1. Observe living and prepared slides of *Euglena* available in the lab, referring to figure 26.8.

26-8

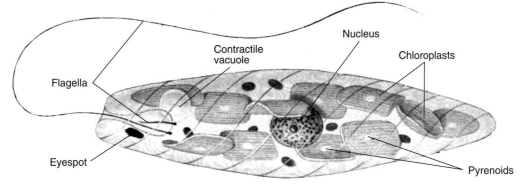

FIGURE 26.8

Euglena, division Euglenophyta. When *Euglena* is in light, starch forms around pyrenoids in this photosynthetic organism.

2. You may want to use a drop of methylcellulose in your preparation to slow the *Euglena*.

3. Note the colored eyespot near the base of the flagella.

Chloroplasts of *Euglena* may contain a single pyrenoid, which appears as a clear, circular area within the plastid. *Euglena* is best known for its ability to be autotrophic, heterotrophic, and saprophytic. Its specific mode of nutrition is determined by current environmental conditions. This phenomenon illustrates why it is often impossible to distinguish plant from animal at the cellular level, and why kingdom Protoctista is necessary. Our classification schemes for these and other organisms will improve as we learn more about them.

Question 11

a. What is the function of the eyespot of *Euglena*? ____

_____

_____

DIVISION MYXOMYCOTA (PLASMODIAL SLIME MOLDS)

The bizarre plasmodial slime molds stream along the damp forest floor in a mass of brightly colored protoplasm called a **plasmodium,** in which individual cells are indistinguishable. Although slime molds are not algae, they are considered members of kingdom Protoctista. Plasmodia are **coenocytic** because their nuclei are not separated by cell walls, and they resemble a moving mass of slime. (The plasmodium of a slime mold should not be confused with the sporozoan genus with the same name.) Plasmodia of slime molds often live beneath detached bark on decomposing tree trunks. Two species occasionally occur on lawns or on mulch beneath shrubs. The moving proto-

plasm of the plasmodia conspicuously pulsates back and forth as they engulf and digest bacteria, yeasts, and other organic particles.

Physarum

Use a dissecting microscope to examine a demonstration culture of *Physarum* growing on oatmeal.

Question 12

a. Is cytoplasmic movement of *Physarum* apparent? ____

_____

b. Is the movement in a particular direction? _____

c. What is a possible function of cytoplasmic

movement in *Physarum*? _____

_____

The *Physarum* on demonstration is in the vegetative plasmodial stage and feeds on the oatmeal. However, if environmental conditions become less than optimal (e.g., if food or moisture decrease), then one of two things can happen. The plasmodium may dry into a hard resistant structure called a **sclerotium** and remain dormant until conditions improve. Or, if light is available, the diploid plasmodium moves to the exposed area and coalesces. The condensed structure grows sporangia, and meiosis produces spores for dispersal (fig. 26.9). Light is associated with an open environment and allows successful reproductive dispersal; dispersal under or within a log would be difficult. Haploid spores produced by meiosis in the sporangia germinate as amoeboid or flagellated organisms. These haploid stages may later fuse as gametes and grow into a new plasmodium.

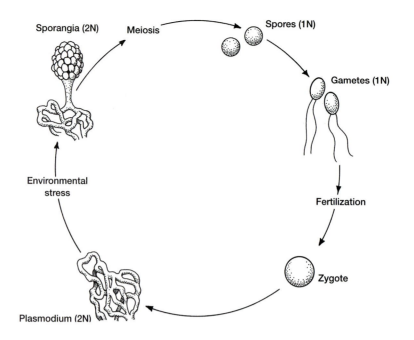

FIGURE 26.9

Life cycle of *Physarum*.

TIRED OF CAMPUS FOOD? READY FOR SOMETHING NEW? TRY THIS!

Some algae are eaten in many parts of the world as part of a normal diet. For example, thin sheets of the red alga *Porphyra* are grown commercially and used in making a Japanese dish called norimake-sushi. Nori is the common name for *Porphyra* in Japan. Try a piece of the dried nori available in lab.

A more palatable way of eating nori is in soup. Here's a recipe:

Nori Soup

    4 cups water
    ½ tsp. sesame oil (optional)
    1 cube chicken or beef bouillon
    2 sheets of nori
    2 eggs
    1 tsp. monosodium glutamate (MSG)
    salt and pepper
    1 or 2 stalks of green onions, chopped

Boil the water and add the bouillon cube and MSG. When the cube is dissolved, add nori and stir. Add well-beaten eggs when nori is soft. Boil for 20 to 30 seconds. Add salt and pepper to taste. Before serving, sprinkle chopped green onions on top of the soup and add the sesame oil. Excellent over steamed rice!

If nori soup doesn't strike your taste, try some "Irish moss," a derivative of the red alga *Chondrus*. Many people eat Irish moss like candy.

Summary of Activities

1. Examine living cultures and prepared slides of *Chlamydomonas*.
2. Examine syngamy of strains of *Chlamydomonas*.
3. Examine living cultures and prepared slides of *Cladophora*, *Spirogyra*, and *Volvox*.
4. Examine preserved slides of *Fucus* and prepared slides of *Fucus* antheridia and oogonia.
5. Examine *Polysiphonia* and *Porphyra*.
6. Examine diatoms and diatomaceous earth.
7. Examine living cultures and prepared slides of dinoflagellates.
8. Examine living cultures and prepared slides of *Euglena*.
9. Examine a demonstration culture of *Physarum*.

Questions for Further Thought and Study

1. Give examples of unicellular, filamentous, and colonial green algae.

2. How do green algae differ from cyanobacteria?

3. What is meant by "alternation of generations"?

4. What is meant by the term "kelp"?

5. Are the stem, holdfast, and blade of brown algae the same as stems, roots, and leaves of land plants? Why or why not?

6. Brown algae contain chlorophyll. Why, then, do they appear brown and not green?

7. What are the main distinguishing features of the major groups of algae?

Doing Botany Yourself

Would you expect environmental conditions to influence syngamy? Design two experiments to investigate the effects of two environmental conditions on the frequency of syngamy. How would you measure syngamy? its frequency? its intensity? its rate?

Writing to Learn Botany

Describe the plantlike and animal-like characteristics of *Euglena*. Which characteristics conclusively define a plant and which ones define an animal?

27

Bryophytes

Objectives

By the end of this exercise you will be able to

1. describe the life histories and related reproductive structures of mosses and liverworts

2. describe the distinguishing features of mosses and liverworts.

The plant kingdom comprises a remarkably diverse group of multicellular organisms. With few exceptions, plants are autotrophic, contain chlorophyll *a*, and have cell walls containing cellulose. The major groups of plants that you will examine in the next four exercises are bryophytes (mosses and liverworts), seedless vascular plants (ferns), gymnosperms (cone-bearing plants), and angiosperms (flowering plants). These groups of plants are distinguished by morphology, life cycle, and the presence or absence of vascular tissues. Pay special attention to variations in each of these characteristics as you survey the plant kingdom in upcoming weeks.

BRYOPHYTES

Bryophytes consist mainly of liverworts, mosses, and hornworts, and they represent the most primitive group of terrestrial plants. Bryophytes are green, have rootlike structures called **rhizoids,** and may have stem and leaflike parts. Bryophytes do not possess vascular tissues, which transport materials between roots and shoots. This absence of vascular tissues in bryophytes typically limits their distribution to moist habitats, since their rhizoids neither penetrate the soil very far nor absorb many nutrients. Also, the absence of vascular tissues necessitates that their photosynthetic and nonphotosynthetic tissues be close together. Because vascular tissues, along with supporting tissues, are absent, bryophytes are relatively small and inconspicuous. Despite their diminutive size, however, bryophytes occur throughout the world in habitats ranging from the tropics to Antarctica. There are approximately 20,600 species of bryophytes, more than any other group of plants except the flowering plants. Bryophytes fix CO_2, degrade rocks to soil, stabilize soil, and reduce erosion. Humans have used bryophytes in a number of ways, including as fuel, to produce Scotch whiskey, and as packing materials.

The plant body of bryophytes is called a **thallus** (pl., thalli). Liverwort thalli are dorsoventrally flattened (from vertical top-to-bottom plane, rather than from side-to-side plane) and bilaterally symmetrical (two equal halves). For comparison, moss thalli are erect and radially symmetrical (circular). Hornwort thalli are similar to those of liverworts.

The life cycle of bryophytes has a distinct alternation of generations in which the gametophyte is the predominant vegetative phase (figs. 27.1 and 27.2). Bryophytes have multicellular sex organs in which gamete-producing cells are enclosed in a jacket of sterile cells. **Antheridia** are male sex organs that produce swimming, biflagellate sperm. Bryophytes require free water for sexual reproduction because their sperm must swim to eggs. These sperm fertilize eggs produced in **archegonia** (sing., archegonium), the female sex organs. The fertilized egg is called a **zygote.** This zygote divides and matures in the archegonium to produce the **sporophyte,** which remains attached to and nutritionally dependent on the **gametophyte.** The mature sporophyte produces haploid spores (via meiosis), each of which can develop into a gametophyte.

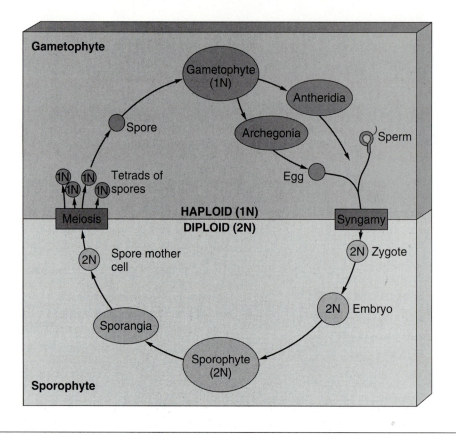

FIGURE 27.1

In a generalized life cycle of a plant, haploid (1N) gametophytes alternate with diploid (2N) sporophytes. Antheridia (male) and archegonia (female), which are the sex organs (gametangia), are produced by the gametophyte; they produce sperm and eggs, respectively. The sperm and egg fuse (syngamy) to produce the first diploid cell of the sporophyte generation, the zygote. Meiosis occurs within the sporangia, the spore-producing organs of the sporophyte. The resulting spores are haploid and are the first cells of the gametophyte generation.

DIVISION HEPATOPHYTA: LIVERWORTS

Although many liverworts appear "leafy," we will restrict our observations to a thallus-type liverwort, *Marchantia*. The gametophytic thallus of this liverwort grows as a large, flat, photosynthetic structure on the surface of the ground (fig. 27.3).

Liverwort Gametophyte

Observe some living *Marchantia*, and note the Y-shaped (dichotomous) growth. Rhizoids extend downward from the lower (ventral) surface of the thallus.

Question 1
a.   What are the functions of rhizoids? _____

_____

View the upper (dorsal) surface of the thallus with a dissecting microscope and note the pores in the center of the diamond-shaped areas. Obtain a prepared slide of a thallus of *Marchantia*, and compare what you see to

figure 27.4. Notice that the pores in the dorsal surface of the thallus overlie air chambers containing **chlorenchyma** (chloroplast-containing) cells.

Question 2
a.   What is the function of the pores in the dorsal

surface of the thallus? _____

_____

Asexual Reproduction in Liverworts

Liverworts can reproduce asexually via fragmentation. In this process, the older, central portions of the thallus die, leaving the growing tips isolated to form individual plants.

Structures called gemmae cups often occur on the dorsal surface of some thalli near the midrib. **Gemmae cups** represent another means of asexual reproduction by liverworts. Inside the gemmae cups are lens-shaped out-growths called **gemmae** (sing., gemma), which are splashed out of the cup by falling drops of rain. If a gemma lands in an adequate environment, it can produce a new gameto-phyte plant. Examine a prepared slide of gemmae cups.

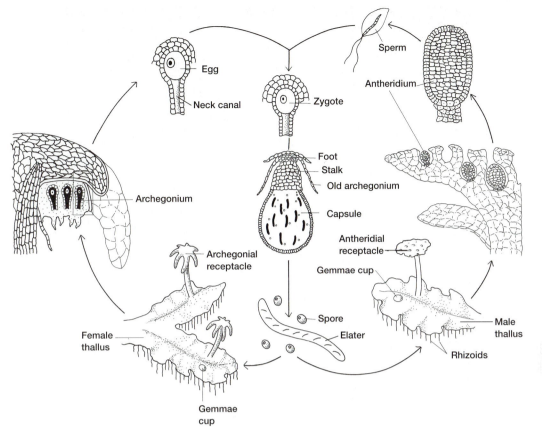

FIGURE 27.2

Life cycle of *Marchantia*, a liverwort. During sexual reproduction, spores produced in the capsule germinate to form independent male and female gametophytes. The archegoniophore produces archegonia, each of which contains an egg; the antheridiophore produces antheridia, each of which produces many sperm. Sperm swim through the neck canal of an archegonium to fertilize the egg. After fertilization, the sporophyte develops within the archegonium and produces a capsule with spores. *Marchantia* reproduces asexually by fragmentation and gemmae.

FIGURE 27.3

Gemmae cups ("splash cups") containing gemmae on the gametophytes of a liverwort. Gemmae are splashed out by raindrops and can then grow into new gametophytes, each identical to the parent plant that produced it by mitosis.

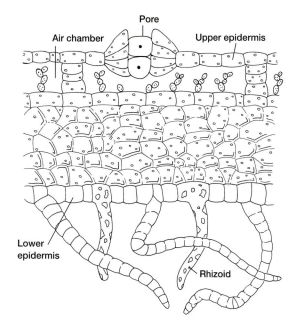

FIGURE 27.4

Thallus of *Marchantia*, a liverwort.

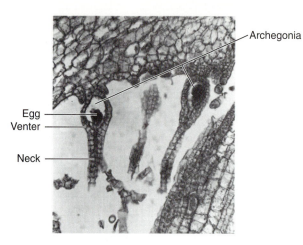

FIGURE 27.5

Marchantia archegonia.

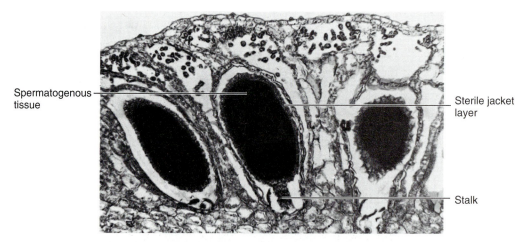

FIGURE 27.6

Marchantia antheridia.

Also examine available live or preserved material. In the space provided here, diagram and label what you see, and compare it to figure 27.3.

Sexual Reproduction in Liverworts

Many species of *Marchantia* are **dioecious,** meaning that they have separate male and female plants. Gametes from each plant are produced in specialized sex organs borne on upright stalks. **Archegoniophores** are specialized stalks on female plants that bear archegonia. Each flask-shaped archegonium consists of a **neck** and a **venter,** which contains the **egg** (fig. 27.5). Examine living or prepared liverworts with mature archegoniophores bearing archegonia. Archegonia at various stages of development are located on the ventral surface. Locate an egg in an archegonium.

Antheridiophores are specialized stalks on male plants that bear antheridia. Examine living or preserved liverworts with mature antheridiophores bearing antheridia. Sperm form in antheridia (fig. 27.6). Flagellated sperm are

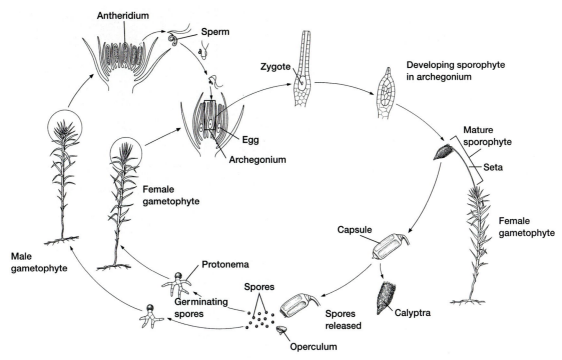

FIGURE 27.7

Moss life cycle. Haploid sperm are released from antheridia on the gametophytes. The sperm then swim through water to the archegonia and down their necks to fertilize the eggs. The resulting zygote develops into a diploid sporophyte. The sporophyte grows out of the archegonium and differentiates into a slender seta with a swollen capsule at its apex. The capsule is covered with a cap, or calyptra, formed from the swollen archegonium. The sporophyte grows on the gametophyte and eventually produces spores after meiosis. The spores are shed from the capsule after a specialized lid—the operculum—drops off. The spores germinate, giving rise to gametophytes. The gametophytes initially are threadlike, they grow along the ground. Ultimately, buds form on them and produce leafy gametophytes.

released from the antheridia and eventually fertilize the egg located in the venter. The zygote remains in the venter and grows into a sporophyte plant.

Examine a prepared slide of cross sections of an antheridiophore. Antheridia are located just below the upper surface of the disk in a chamber that leads to the surface of the disk through a pore.

Question 3

a. How do the positions of the archegonium and antheridium relate to their reproductive functions?

_____

_____

Liverwort Sporophyte

Examine a prepared slide of a sporophyte of *Marchantia*. The nonphotosynthetic sporophyte is connected to the gametophyte by a structure called the **foot**. **Spores** are produced by meiosis in a **capsule** located on a **stalk** that extends downward from the foot. Among the spores you can see elongate cells called **elaters**. Elaters help disperse spores by twisting. In humid conditions the elaters coil, but when it is dry the elaters expand, pushing the spores apart and rupturing the spore case to release the spores.

Question 4

a. What is the function of the foot? _____

_____

b. Are the spores haploid or diploid? _____

c. Do these spores germinate into a gametophyte or sporophyte? _____

d. What is the functional significance of the response of elaters to moisture? _____

_____

DIVISION BRYOPHYTA: MOSSES

Mosses are seen more frequently than liverworts because of their greater numbers, more widespread distribution, and the fact that gametophyte plants of mosses are leafy and usually stand upright. Mosses also withstand desiccation better than liverworts. Consequently, mosses have less specialized habitats. The moss gametophyte is radially symmetrical and is the most conspicuous phase of the moss life cycle (fig. 27.7).

Moss Gametophyte

Observe the living moss on display called *Polytrichum* (fig. 27.8). The "leafy" green portions of the moss are the gametophytes. Make a wet mount of a single leaflet and examine it with low magnification.

Question 5

a. How many cells thick is the leaflet? _____

b. Is there a midrib? _____

   a vein? _____

c. Are stomata or pores visible on the leaf surface?

   _____

d. How does the symmetry of a moss gametophyte compare with that of a liverwort gametophyte?

   _____

   _____

Moss gametophytes have specialized cells that aid in the absorption and retention of water. Mats of moss act, in effect, like sponges. The following exercise demonstrates the water-absorbing potential of mosses.

Procedure 27.1
Water absorption by moss

1. Weigh 3 g of whole *Sphagnum*, a peat moss, and 3 g of paper towel. Do not use powdered or ground moss.

2. Add the moss and towel to separate beakers each containing 100 mL of water.

3. After several minutes, remove the materials from the beaker.

4. Measure the amount of water left in each beaker by pouring the water into a 100 mL graduated cylinder. Remember that 1 mL of water weighs 1 g.

5. Record your data.

Question 6

a. How many times its own weight did the moss absorb?

   _____

b. How does this compare with the paper towel? _____

   _____

c. Why is *Sphagnum* often used in shipping items that

   must be kept moist? _____

   _____

Asexual Reproduction in Mosses

Unlike liverworts, mosses lack structures such as gemmae for asexual reproduction. Mosses reproduce asexually by fragmentation.

FIGURE 27.8

Male gametophyte of *Polytrichum*, a moss.

Sexual Reproduction in Mosses

Most mosses, like liverworts, are dioecious. Archegonia or antheridia are borne either on tips of the erect gametophyte stalks or as lateral branches on the stalks. The apex of stalks of the female plant (the plant that bears archegonia) appears as a cluster of leaves, with the archegonia buried inside. Examine living or preserved mosses with mature archegonia.

Examine a prepared slide of moss archegonia (fig. 27.9). Note the canal that leads through the neck and terminates in the venter of the archegonium. When the archegonium matures, cells lining the neck disintegrate and form a canal leading to the egg. Sperm, following a chemical attractant released by the archegonium, swim through this canal to reach the egg. Try to relate what you see in the cross section to the whole plant.

Question 7

a. Where is the egg located in the archegonium?

   _____

The male plant (the plant that bears antheridia) has a platelike structure on the tip with the "leaves" expanding outward to form a rosette. This structure is sometimes

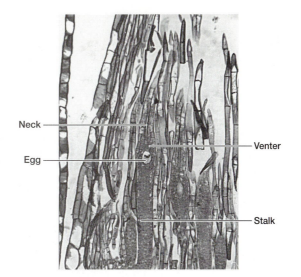

FIGURE 27.9

Polytrichum archegonia.

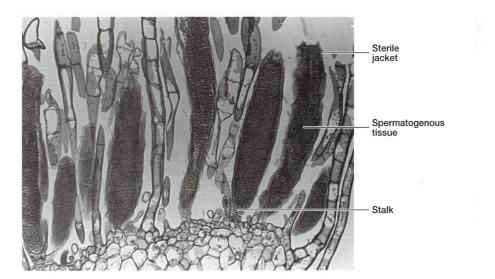

FIGURE 27.10

Polytrichum antheridia.

called a "moss flower" because of its appearance, or a "splash cup" because of its function (i.e., the dispersal of sperm by falling raindrops). Examine living or preserved mosses with mature antheridia.

Examine a prepared slide of moss antheridia, which appear as elongate, saclike structures (fig. 27.10). Locate the outer sterile jacket and the inner mass of cells destined to become sperm.

Moss Sporophyte

Moss sporophytes consist of **capsules** located atop stalks, or **seta,** that extend upward from the moss gametophyte. A sporophyte is attached to the gametophyte by a structure called a foot.

Question 8

a. Is the sporophyte more prominent in mosses or liverworts? _____

The capsule atop the seta is covered by the **calyptra,** which is the upper portion of archegonium that covers the apex of the capsule. The calyptra falls off when the capsule matures. Inside the capsule are numerous haploid spores formed via meiosis.

Question 9

a. What is the adaptive significance of having spores released from atop a seta? _____

_____

If enough living material is available in the lab, remove the calyptra from a sporophyte capsule. On the tip of the capsule is a caplike structure called the **operculum.** Remove the operculum and notice the hairlike teeth lining the opening of the capsule. These teeth help control the release of spores from the capsule. In wet weather these teeth bend inward and prevent release of the spores. In dry weather they bend outward, facilitating distribution of spores by the wind.

Crush the capsule in water on a microscope slide, and note the large number of spores that are released.

Question 10

a. What process produces spores? _____

b. Is the sporophytic capsule haploid or diploid? _____

Moss spores germinate and form a photosynthetic protonema, which resembles a branching, filamentous alga. Leafy moss plants arise from "buds" located along the protonema.

Question 11

a. Can you think of any evolutionary implications of the similarity between a moss protonema and a

filamentous green alga? _____

_____

DIVISION ANTHOCEROTOPHYTA: HORNWORTS

The hornworts are the smallest group of bryophytes; there are only about 100 species in six genera. Hornworts have several features that distinguish them from most other bryophytes. The sporophyte is shaped like a long, tapered horn that protrudes from a flattened thallus. Archegonia are not discrete organs. Rather, they are embedded in the thallus and are in contact with surrounding vegetative cells.

The most familiar hornwort is *Anthoceros,* a temperate genus. Examine some living and preserved *Anthoceros.* Locate the gametophytic thallus and the rhizoids extending from its lower surface. Also locate the hornlike sporophytes extending from the upper surface. The spores are produced in the horn of the sporophyte. If prepared slides are available, locate spores in various stages of development within the sporophyte.

Summary of Activities

1. Examine the general life cycle of bryophytes.
2. Examine living *Marchantia* and compare it to a prepared cross section of a thallus.
3. Examine a prepared slide of gemmae cups.
4. Examine live or prepared liverworts with mature archegoniophores bearing archegonia and antheridiophores bearing antheridia.

5. Examine a prepared slide of a sporophyte of *Marchantia.*
6. Observe living mosses.
7. Demonstrate water absorption by *Sphagnum.*
8. Examine prepared slides of moss archegonia and antheridia.
9. Examine capsules of a living moss sporophyte.
10. Examine living *Anthoceros* and prepared slides of *Anthoceros* sporophyte.

Questions for Further Thought and Study

1. List advances in complexity of bryophytes over algae regarding their morphology, habitat, asexual reproduction, and sexual reproduction.
2. What event begins the sporophyte phase of the life cycle? Where does it occur in mosses and liverworts?
3. What event begins the gametophyte phase of the life cycle? Where does it occur in mosses and liverworts?
4. What features distinguish a moss from a liverwort?
5. Diagram the life cycle of a liverwort, indicating which stages are sporophytic and which are gametophytic.
6. Diagram the life cycle of a moss, indicating which stages are sporophytic and which are gametophytic.
7. What ecological roles do mosses, liverworts, and hornworts play in the environment?
8. Is the sporophyte of mosses ever independent of the gametophyte? Explain.
9. Why do you think that bryophytes are sometimes referred to as the amphibians of the plant kingdom?
10. How did liverworts obtain their name?

Doing Botany Yourself

Mosses and liverworts are surprisingly diverse. Make a simple collection of living mosses and liverworts from two or three sites. Bring the collection to class and compare it to those of other students. Use reference books in your lab or library to identify the plants that you have collected.

Writing to Learn Botany

Since water is required for the swimming sperm to reach the archegonium, would you say that bryophytes are not truly land plants? Why or why not?

Seedless Vascular Plants

Objectives

By the end of this exercise you will be able to

1. discuss the similarities and differences between ferns and other plants you have studied
2. describe the life cycles of seedless vascular plants
3. describe the distinguishing features of Psilotophyta (whisk ferns), Equisetophyta (horsetails), Lycopodophyta (club mosses), and Pteridophyta (ferns).

Seedless vascular plants include nonflowering plants that have a vascular system of fluid-conducting xylem and phloem. The vascular system connects the leaves, roots, and stems.

Seedless vascular plants include the divisions Pteridophyta (ferns), Lycopodophyta (club mosses), Psilotophyta (whisk ferns), and Equisetophyta (horsetails). Biologists often have referred to these divisions as the lower vascular plants or ferns and their allies. Conversely, gymnosperms (e.g., pine trees) and angiosperms (flowering plants) have been referred to as higher vascular plants or seed-forming vascular plants. This separation of higher and lower vascular plants was invented by early botanists, who believed that all seed-forming plants must be related, and that plants lacking seeds are similarly related.

Later studies of lower vascular plants showed that ferns represent a separate group of seedless plants distinguished by the presence of **megaphylls,** which are large leaves with several to many veins. The remaining seedless vascular plants possess **microphylls** (small leaves with one vein) and are only somewhat related to ferns; hence the name "fern allies." Thus, the lower vascular plants include a diverse group of divisions. However, all seedless vascular plants possess **sporophylls,** which are leaflike structures of the sporophyte generation that bear spores. Sporophylls may be large and have several to many veins (as in the megaphylls of ferns), or they may be smaller and have one vein (as in the microphylls of the other divisions of seedless vascular plants).

DIVISION PTERIDOPHYTA: FERNS

Ferns inhabit almost all kinds of environments and possess characteristics of the more advanced seed plants as well as the less advanced bryophytes. Ferns have an independent sporophyte (fig. 28.1) with well-developed vascular tissue. The diversity of ferns is striking; they range from majestic tree ferns to bizarre staghorn ferns. Tree ferns reach heights of up to 16 meters. Along with other plants, these ferns once formed forests that were transformed into coal deposits. Today, humans use ferns as decorations, fossil fuels, and in rice cultivation. Review the fern life cycle shown in figure 28.1 and plate VIII, fig. 15.

Question 1
a. Which parts of the life cycle of ferns are haploid?

_____

_____

b. Which are diploid? _____

_____

Fern sporophytes grow indefinitely via underground stems called rhizomes. Examine the fern **rhizomes** on display. Examine the different ferns available in the lab and note the different shapes of the leaflike **fronds.** Identify the **stalk, blade,** and **pinnae.**

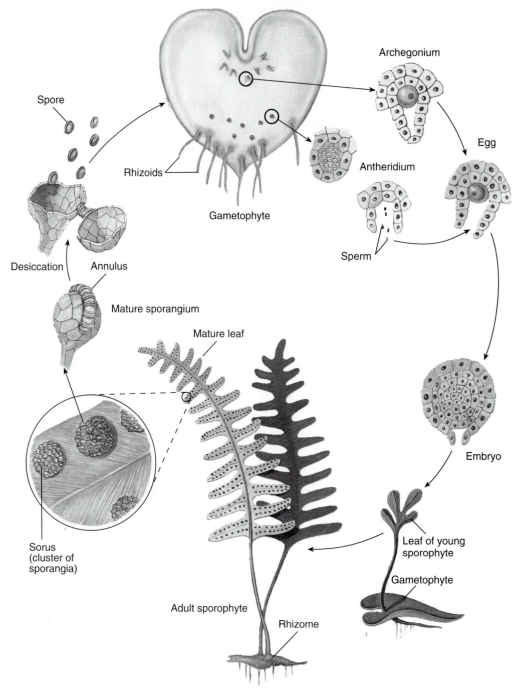

Spore

Archegonium

Rhizoids

Antheridium

Gametophyte

Egg

Desiccation   Annulus

Sperm

Mature sporangium

Mature leaf

Embryo

Sorus
(cluster of
sporangia)

Leaf of young
sporophyte

Gametophyte

Adult sporophyte

Rhizome

FIGURE 28.1

Fern life cycle. The haploid gametophytes grow in moist places. Rhizoids (anchoring structures) project from their lower surface. Eggs and sperm develop in archegonia and antheridia, respectively, on gametophytes' lower surface near the apical notch. Released sperm swim through free water to the archegonia, entering and fertilizing the single eggs. The zygote—the first cell of the diploid sporophyte generation—starts to grow within the archegonium. Eventually, the sporophyte becomes much larger than the gametophyte. Most ferns have more or less horizontal stems, called rhizomes, that creep along below the ground. On the sporophyte's leaves, called fronds, are clusters of sporangia within which meiosis occurs and spores are formed. The released spores germinate and become new gametophytes.

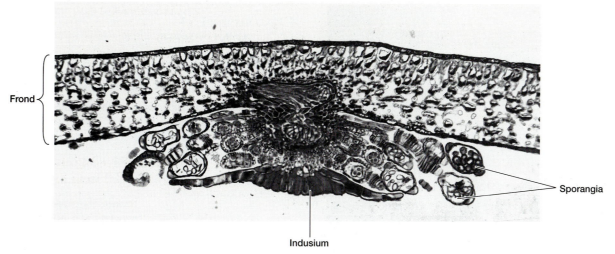

Front {

Sporangia

Indusium

FIGURE 28.2

Fern sorus.

Question 2

a. How many veins are present in each frond? _____

b. What tissues compose a vein? _____

_____

c. What is the function of the stalk? _____

the blade? _____

the pinnae? _____

Groups of sporangia called **sori** form on the underside of fern fronds. **Sporangia** may be protected by a shield-shaped **indusium,** which is a specialized outgrowth of the frond. Meiosis in the sporangium produces haploid spores, which are the first stage of the gametophyte.

Procedure 28.1
Examine sori

1. Scrape a sorus into a drop of water and observe it using low power on your microscope. You'll notice a row of thick-walled cells along the back of the helmet-shaped sporangium. These cells are the annulus.

Question 3

a. What is the function of the annulus? _____

_____

b. Are any spores in the sporangium? _____

2. Place a few drops of acetone on the sporangium while you observe it with a dissecting microscope.

3. Watch the sporangium for a few minutes, adding acetone as needed.

4. Describe what you see.

Question 4

a. Did the application of acetone cause the spores of

the fern to disperse? _____

b. How is the mechanism for spore dispersal in ferns

similar to that of bryophytes? _____

5. Examine prepared slides of fern sori, referring to figure 28.2.

6. Diagram and label each structure that you see, and list its function.

Fern Reproduction

Fern spores germinate and form a threadlike **protonema.** Subsequent cellular divisions produce an independent, heart-shaped **prothallium** ("valentine plant").

Question 5

a. Is the prothallium haploid or diploid? _____

b. Is the prothallium sporophyte or gametophyte?

_____

Rhizoids and reproductive structures occur on the underside of the prothallium. Globe-shaped antheridia form first, followed by archegonia. After producing sperm, antheridia drop off, leaving sperm to swim to the archegonia of other prothallia and fertilize the eggs. Vase-shaped archegonia contain eggs and are located near the cleft of the heart-shaped prothallium. Observe

FIGURE 28.3

Lycopodium, a club moss.

archegonia and antheridia (see fig. 28.1) on prepared slides and on a living prothallium.

Question 6
a.   What is the adaptive significance of having these structures on the lower surface of the prothallium rather than on the upper surface? _____

_____

b.   What is the adaptive significance of having sperm and egg produced at different times? _____

_____

The zygote develops in the archegonium and is nutritionally dependent on the gametophyte for a short time. Soon thereafter, the sporophyte becomes leaflike and crushes the prothallium. Fronds of the growing sporophyte break through the soil in a coiled position called a **fiddlehead.** The fiddlehead then unrolls to display the frond, which is a single leaf. Fiddleheads are considered a culinary delicacy in some parts of the world.

Most terrestrial ferns are **homosporous;** they produce one kind of spore, which develops into a single kind of gametophyte, which produces both antheridia and archegonia. Conversely, aquatic ferns such as *Salvinia* and *Azolla* are **heterosporous,** meaning that they produce two kinds of spores: **megaspores** and **microspores.** A megaspore forms a gametophyte with only archegonia, and a microspore forms a gametophyte with only antheridia. Observe living *Salvinia* and *Azolla* in the lab.

Question 7
a.   How do *Salvinia* and *Azolla* differ from other ferns you examined earlier? _____

_____

DIVISION LYCOPODOPHYTA (CLUB MOSSES)

Club mosses have true roots, stems, and leaves. Most asexual reproduction by club mosses occurs via rhizomes. If available, locate the rhizome on a *Lycopodium,* a club moss. Study the aboveground portion of the *Lycopodium* plant (fig. 28.3). *Lycopodium* is evergreen, as are some ferns, most gymnosperms, and some angiosperms.

Question 8
a.   How could a rhizome be involved in asexual

reproduction? _____

_____

b.   How is a rhizome different from a rhizoid? _____

_____

c.   Do rhizomes have leaves? _____

d.   What are the shape and size of the leaves? _____

_____

e.   What is the significance of the form of the leaves?

_____

_____

f.   Is a midvein visible? _____

g.   What does the term "evergreen" mean? _____

_____

h.   Is being evergreen a good characteristic for

classifying plants? _____

Why or why not? _____

_____

Sporangia of *Lycopodium* occur on small modified leaves called **sporophylls,** which are clustered in strobili (cones) that form at tips of branches. Examine strobili on a living *Lycopodium* plant. Examine prepared slides of strobili of *Lycopodium* (fig. 28.4). Diagram, label, and state the function of each major feature of the strobilus.

Examine prepared slides of spores of *Lycopodium* as well as those available on plants growing in the lab.

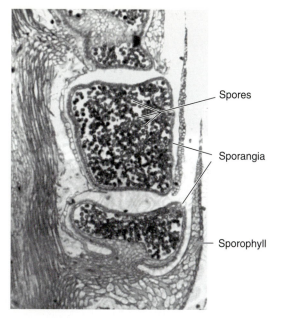

FIGURE 28.4

Lycopodium sporophyll and sporangium.

Question 9

a. How many sporangia occur on each sporophyll of
Lycopodium? _____

b. Is *Lycopodium* homosporous or heterosporous?

_____

c. Can you see why spores of *Lycopodium* are sometimes
called "vegetable sulfur"? _____

d. Why are the spores a good dry lubricant? _____

_____

e. Which is the dominant part of the *Lycopodium* life
cycle, the sporophyte or gametophyte? _____

Examine some living *Selaginella* plants (fig. 28.5).
Many species of *Selaginella* produce two kinds of strobili,
which are typically red and yellow. If strobili are present,
examine spores derived from these cones. Examine
hydrated and dehydrated resurrection plants (*Selaginella
lepidophylla*).

Question 10

a. Are spores of *Selaginella* similar in size? _____

b. What is this condition called? _____

c. What is the functional significance of the difference
in the appearance of dehydrated and rehydrated
Selaginella? _____

_____

FIGURE 28.5

Selaginella.

d. Can you see why these plants are sometimes referred
to as "resurrection plants"? _____

Examine some living *Isoetes* (quillwort) plants. Compare and contrast the following features of *Isoetes* with *Lycopodium* and *Selaginella:*

- Shape of aerial portion of plant
- Branching patterns
- Shape, size, and arrangement of leaves

At the branching points along the stem of *Isoetes*, you'll
see an unusual runnerlike organ. These proplike axes are
called **rhizophores** and have structural features that are intermediate between stems and roots.

FIGURE 28.6

Psilotum.

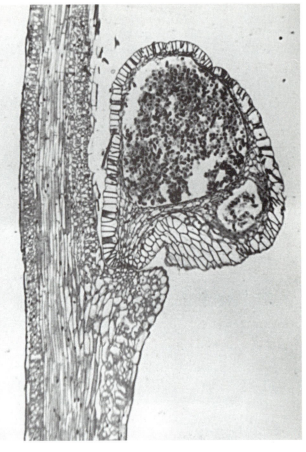

FIGURE 28.7

Psilotum sporangium.

DIVISION PSILOTOPHYTA: WHISK FERNS

There are only two extant representatives of the division Psilotophyta: *Psilotum* (fig. 28.6) and *Tmesipteris*. *Psilotum* has a widespread distribution, while *Tmesipteris* is restricted to the South Pacific. *Psilotum* lacks leaves and roots and is homosporous. Examine a prepared slide of a *Psilotum* sporangium (fig. 28.7), and study living *Psilotum* plants in the lab.

Question 11
a.   What type of branching characterizes *Psilotum?*

_____

b.   Are any roots present? _____

c.   Are any leaves present? _____

d.   Where are the sporangia? _____

e.   Where does photosynthesis occur in *Psilotum?* _____

_____

DIVISION EQUISETOPHYTA: HORSETAILS

Equisetum, the only extant member of Equisetophyta, is an example of a plant whose vegetative structure identifies the plant better than does its reproductive structure. *Equisetum* is distinguished from other fern allies by its jointed and ribbed stem. Examine the living *Equisetum* plants.

Question 12
a.   Where are the leaves? _____

b.   What part of the plant is photosynthetic? _____

c.   Which part of the life cycle of *Equisetum* is dominant,

the sporophyte or gametophyte? _____

Feel the ribbed stem of an *Equisetum* plant. Its rough texture results from siliceous deposits in its epidermal cells. In frontier times, *Equisetum* was used to clean pots and pans, sand wooden floors, and scour plowshares, thus accounting for its common name of "scouring rush."

Strobili of *Equisetum* occur at the tips of reproductive stems. Sporangia form atop umbrella-like structures called **sporangiophores,** which are modified branches. Elaters in sporangia of *Equisetum* help disperse spores.

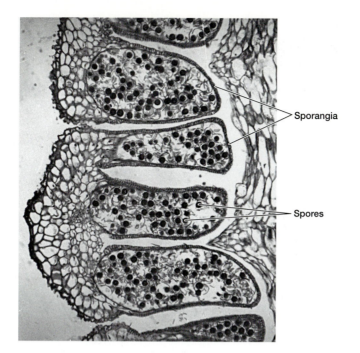

FIGURE 28.8

Equisetum strobilus with sporangia.

Examine prepared slides of *Equisetum* strobili (fig. 28.8). Diagram, label, and state the function of each major structure comprising the strobilus.

Question 13

a. How does strobili formation in *Equisetum* compare with *Lycopodium* and *Selaginella*? _____

_____

b. How is spore dispersal aided by elaters? _____

_____

To review the structures and characteristics of seedless vascular plants, complete table 28.1.

Summary of Activities

1. Review the life cycle of ferns.
2. Examine sori.
3. Examine fern archegonia and antheridia on prepared slides.
4. Examine living prothallia.
5. Observe living *Salvinia* and *Azolla*.
6. Examine the aboveground tissue, rhizome, sporophylls, strobili, and spores of *Lycopodium*, a club moss.
7. Examine hydrated and dehydrated *Selaginella*.
8. Examine *Psilotum* (whisk fern) and *Isoetes* (quillwort).
9. Examine living *Equisetum* and prepared slides of strobili.
10. Compare the function of structures common to seedless vascular plants.

TABLE 28.1

SUMMARY OF STRUCTURES COMMON TO SEEDLESS VASCULAR PLANTS

Plant Structure	Sporophyte or Gametophyte	Function
Prothallium		
Pinnae		
Spore		
Frond		
Annulus		
Sporangium		
Antheridium		
Archegonium		
Microspore		
Megaspore		
Microphyll		
Megaphyll		

Questions for Further Thought and Study

1. Compare ferns with bryophytes. Which represent the best "engineering job"? Explain.
2. What are the distinguishing features of horsetails, whisk ferns, and club mosses? How are they different from ferns?
3. What are the advantages of vascular tissue in land plants?

Writing to Learn Botany

What problems were faced by plants as they moved onto land? What adaptations of mosses, liverworts, and ferns are relevant to the transition?

29

Gymnosperms

Objectives

By the end of this exercise you will be able to

1. describe the distinguishing features of gymnosperms
2. understand the life cycle of pine, a representative gymnosperm
3. understand some adaptations of pine to cold, dry environments
4. identify the parts and understand the function of a cone
5. identify the parts and understand the function of a seed.

Gymnosperms are plants with exposed seeds borne on scalelike structures called **cones** (strobili). Like ferns, gymnosperms have a well-developed alternation of generations. Unlike ferns, however, gymnosperms are **heterosporous,** that is, they produce two kinds of spores (fig. 29.1). **Microspores** occur in male cones and form male gametophytes. **Megaspores** occur in female cones and form female gametophytes. Gametophytes of gymnosperms are microscopic and completely dependent on the large, free-living sporophyte. Gymnosperms include four divisions: Cycadophyta, Ginkgophyta, Pinophyta, and Gnetophyta. The last of these, the Gnetophyta, is not discussed here because the division includes only a few rare genera.

DIVISION CYCADOPHYTA

The Cycadophyta (cycads) once flourished, but today they consist of only about ten genera and 100 species. Cycads resemble palms because they have unbranched trunks and large, closely packed leaves that are evergreen and tough (fig. 29.2). Sperm of cycads are flagellated.

Examine a branch bearing the developing seeds of a cycad. Note that the seeds are fleshy and exposed to the environment.

Question 1

a. What is the evolutionary significance of flagellated sperm in cycads? _____

_____

DIVISION GINKGOPHYTA

The ginkgophyta consists of one species, *Ginkgo bilboa* (Maidenhair plant), a large dioecious tree that does not bear cones. *Ginkgo* are hardy plants in urban environments and tolerate insects, fungi, and pollutants. Males are usually planted because females produce fleshy, smelly, and messy fruit that superficially resemble cherries. Leaves of *Ginkgo* have a unique, shape (fig. 29.3). *Ginkgo* has not been found in the wild and would probably be extinct but for its cultivation in ancient Chinese and Japanese gardens.

Question 2

a. What does dioecious mean? _____

_____

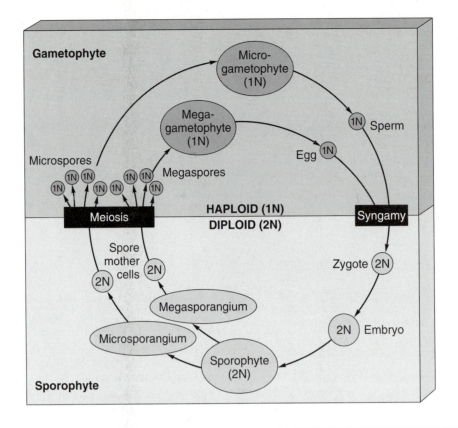

FIGURE 29.1

Diagram of the life cycle of a heterosporous vascular plant. The diploid sporophyte alternates with the haploid gametophyte stage. The gametophyte generation occurs as separate microgametophytes and megagametophytes that produce sperm and egg nuclei, respectively.

FIGURE 29.2

Cycad. A cone of exposed seeds forms among a rosette of leaves.

FIGURE 29.3

Ginkgo.

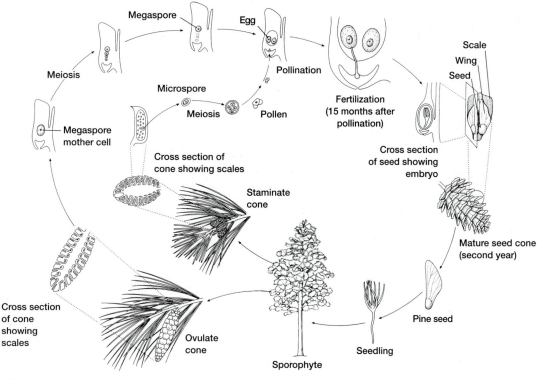

FIGURE 29.4

Pine life cycle. In seed plants, the gametophyte generation is greatly reduced. A germinating pollen grain is the mature microgametophyte of a pine. Pine microsporangia are borne in pairs on the scales of the delicate pollen-bearing cones. Megagametophytes, in contrast, develop within the ovule. The familiar seed-bearing cones of pines are much heavier structures than the pollen-bearing cones. Two ovules, and ultimately two seeds, are borne on the upper surface of each scale. In the spring, when the seed-bearing cones are small and young, their scales are slightly separated. Drops of sticky fluid form between these scales and capture airborne pollen. These pollen grains germinate, and slender pollen tubes grow toward the egg. When a pollen tube grows to the vicinity of the megagametophyte, sperm are released, fertilizing the egg and producing a zygote there. The development of the zygote into an embryo occurs within the ovule, which matures into a seed. Eventually, the seed falls from the cone and germinates, the embryo resuming growth and becoming a new pine tree.

DIVISION PINOPHYTA

The Pinophyta (conifers) are a large group of cone-bearing plants that includes the 5,000-year-old bristlecone pines, probably the earth's oldest living individual organisms. Cones are reproductive structures of the sporophyte generation that consist of several scalelike sporophylls arranged about a central axis. Sporophylls, also present in ferns and their allies, are modified leaves specialized for reproduction. Specifically, sporophylls bear spores. In conifers, sporophylls of male cones are called **microsporophylls.** On the surface of each microsporophyll is a layer of cells called a **microsporangium** that produces spores. Sporophylls of female cones are **megasporophylls,** and each bears two spore-producing **megasporangia** on its upper surface. Both microsporangia and megasporangia are patches of cells near the central axis of the sporophylls composing the cones on the respective sporophylls. Male cones are small and similar in all conifers. However, female cones are variable; they may be small (1–3 mm in *Podocarpus*) and fleshy (a *Juniperus* "berry") or large and woody (6–40 cm in *Pinus*).

In this exercise you'll study pine (*Pinus*), a representative and widely distributed conifer. Pine has considerable economic value because it is used to produce lumber, wood pulp, pine tar, resin, and turpentine. Other examples of conifers include spruce, cedar, and fir.

The Pine Sporophyte

The life cycle of *Pinus* is typical of conifers (fig. 29.4). You are already familiar with the sporophyte of pine—it is the tree. Examine pine twigs with leaves (needles) and a terminal bud. Notice that the leaves are borne on dwarf branches only a few millimeters long. The length and number of leaves distinguishes many of the species of *Pinus*. Examine a prepared slide of a pine-leaf cross section and locate the structures labeled in figure 29.5.

Question 3
a. Where on the branch was the terminal bud last year?

_____

b. How can you tell? _____

_____

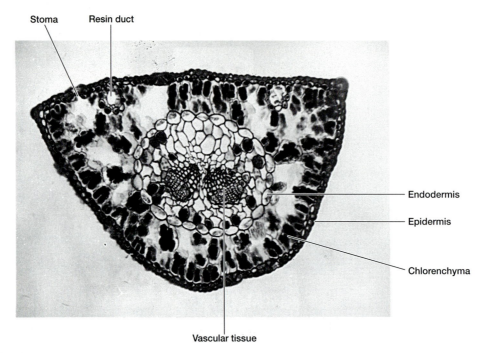

Stoma   Resin duct

Endodermis

Epidermis

Chlorenchyma

Vascular tissue

FIGURE 29.5

Pine leaf, cross section.

c. How are needles or leaves arranged? _____

_____

d. How many leaves are in a bundle? _____

e. How are pine leaves different from those of

deciduous plants? _____

_____

f. Why are pines called evergreens? _____

_____

g. What other plants have you studied that are

evergreen? _____

_____

h. What is the function of each of the structures labeled

in figure 29.5? _____

_____

i. How do the structural features of pine leaves adapt

the tree for life in cold, dry environments? _____

_____

Observe pine branches with staminate (male) and ovulate (female) cones. Examine a prepared slide of a young staminate cone and note the pine pollen in various stages of development. Each scale (microsporophyll) of the male cone bears a microsporangium, which in turn produces diploid **microspore mother cells** (microsporocytes)

(fig. 29.6). Microspore mother cells undergo meiosis to produce **microspores** that develop into microgametophytes called **pollen grains** (fig. 29.7). Each pollen grain consists of a generative nucleus that divides to form a stalk nucleus (vestige of an antheridium), a body nucleus that later divides to form two sperm nuclei, and a pair of bladderlike wings. Note that the gametophytes of gymnosperms are reduced to only a few nuclei.

Prepare a wet mount of some pine pollen; note their characteristic shape. Also, remove a scale from a mature staminate cone and tease open the microsporangium. Prepare a wet mount of the microsporangium and its contents, and examine the contents with your microscope. Draw the shape of a pine pollen grain below.

Question 4

a. Are all of the cones the same size? _____

b. How might the arrangement and location of

staminate cones help ensure cross-pollination?

_____

_____

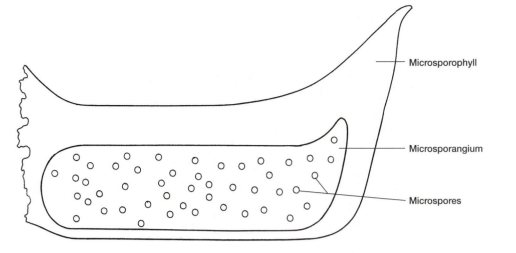

Microsporophyll

Microsporangium

Microspores

FIGURE 29.6

Microsporophyll with microsporangium and microspores.

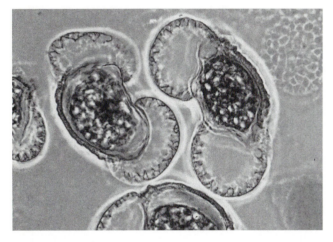

FIGURE 29.7

Pollen grains of *Pinus*.

(a)

c.   What is the probable function of the wings of a pine

pollen grain? _____

_____

Examine young living or preserved ovulate cones (fig. 29.8). These cones develop and enlarge considerably before they are mature. Examine a prepared slide of a young ovulate cone ready for pollination. Each ovuliferous scale of the female cone bears two megasporangia, each of which produces a diploid megaspore mother cell. Each **megaspore mother cell** (megasporocyte) undergoes meiosis to produce a **megaspore,** which develops into a **megagametophyte** (free-nuclear female gametophyte). The tissue of the megasporangium immediately surrounding the megagametophyte is the **nucellus,** which is surrounded by **integuments.** A megagametophyte and its surrounding tissues constitute an **ovule** and contain at least one archegonium

(b)

FIGURE 29.8

Pine cones. (*a*) First-year ovulate (seed) cones open for pollination. (*b*) Second-year pine cones at time of fertilization.

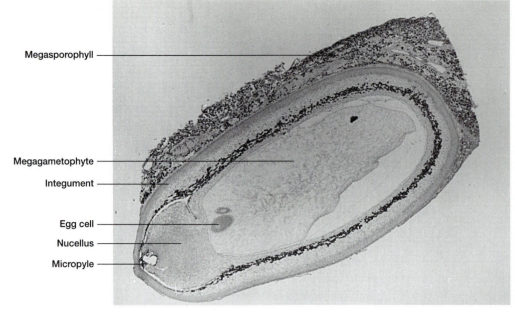

Megasporophyll ———

Megagametophyte ———
Integument ———

Egg cell ———
Nucellus ———
Micropyle ———

FIGURE 29.9

Pine ovule section.

with an egg cell. Examine a prepared slide of an ovulate cone that has been sectioned through an ovule (fig. 29.9). An ovule develops into a seed.

Examine a mature ovulate cone and notice its spirally arranged **ovuliferous scales.** These scales are analogous to microsporophylls of staminate cones, but ovuliferous scales are modified branches rather than modified leaves. At the base of each scale you'll find two naked seeds. Notice that the seeds are exposed to the environment and are supported (but not covered) by an ovuliferous scale.

Question 5
a.   On which surface of the scale are the seeds located?

_____

b.   How large is a staminate cone compared to a newly

pollinated ovulate cone? _____

a mature ovulate cone? _____

Pollination and Seed Formation in Pine

Pollination is the transfer of pollen to a receptive surface. Pollen grains released in late spring are carried by wind to ovulate cones, in which pollen sifts through the ovuliferous scales and sticks in a drop of resin at the micropylar end of an ovule. As the resin dries, pollen is drawn through an opening in the ovule called the micropyle and into contact with the ovule. The pollen grain then germinates and grows a tube into the archegonium, where it releases its two nonmotile sperm nuclei. One of these nuclei disintegrates; the

other fuses with the egg to form a zygote. The zygote, while in the ovule, develops into the embryo of a new sporophyte. The ovule is now a seed and consists of an embryo, a seed coat (integuments of the megasporangia), and a food supply (tissue of the megagametophyte).

Examine a prepared slide of a pine seed. Locate the embryo, seed coat, and food supply. Seeds are released when the cone dries and the scales separate. This usually occurs within 15 months after pollination. Examine some mature pine seeds, noting the winglike extensions of the seed coat.

Seed

The evolution of seeds is one of the most significant events in the history of the plant kingdom. Indeed, the evolution of seeds is one of the factors responsible for the dominance of seed plants in today's environment, because a seed permits a small but multicellular sporophyte to remain dormant until conditions are favorable for continued growth. While dormant, the young, embryonic sporophyte is protected by a seed coat and surrounded by a food supply.

Refer to the pine life cycle shown in figure 29.4 to answer question 6. If plant material and time are available, examine other gymnosperms in the lab.

Question 6
a.   Where are spores located? _____

b.   What is the male gametophyte? _____

_____

c. What is the female gametophyte? _____

_____

d. What is an ovule? _____

_____

e. What is an integument? _____

_____

f. What is the function of the winglike extensions of a pine seed? _____

_____

g. How are other gymnosperms similar to pine? _____

_____

h. How are they different? _____

_____

DIVISION GNETOPHYTA

The gnetophytes (seventy-one species in three genera) include some of the most distinctive (if not bizarre) of all seed plants. They have many similarities with angiosperms such as flowerlike compound strobili, vessels in the secondary xylem, loss of archegonia, and double fertilization. The slow growing *Welwitschia* plants are extraordinary in appearance, and they live only in the deserts of southwestern Africa. Most of its moisture is derived from fog that rolls in from the ocean at night. *Welwitschia* stems are short and broad (1 m). Mature, 100-year-old plants have only two leaves that are wide and straplike.

Summary of Activities

1. Examine a branch and seeds of a cycad.
2. Examine pine twigs with leaves and a terminal bud.
3. Examine a prepared slide of a cross section of a pine leaf.
4. Examine pine branches with male and female cones.
5. Examine a prepared slide of a young staminate cone.
6. Prepare a wet mount of pine pollen.
7. Examine a mature ovulate cone.
8. Examine a prepared slide of an ovulate cone ready for pollination.
9. Examine an ovulate cone sectioned through an ovule.
10. Examine a prepared slide of a pine seed as well as whole pine seeds.
11. Examine additional gymnosperms available in the lab.

Questions for Further Thought and Study

1. What is the difference between pollination and fertilization?
2. How is alternation of generations different in ferns and pines?
3. How are the environmental agents for uniting sperm and egg different in pines and bryophytes?
4. What does the term "gymnosperm" mean?

Writing to Learn Botany

What ecological advantages does the seed habit have over the reproductive process in bryophytes?

30

Angiosperms

DIVISION MAGNOLIOPHTA

Flowering plants (division Magnoliophyta, also referred to as angiosperms) are the most abundant, diverse, and widespread of all land plants. They owe their success to several factors: their structural diversity, efficient vascular systems, mutualisms (especially with fungi and insects), and short generation times (see plates IX–XVI). Angiosperms range in size from 1 mm (*Wolffia*) to over 100 m tall (*Eucalyptus*). Like gymnosperms, the sporophyte of angiosperms is large and independent of the microscopic gametophyte. This is opposite of the situation in bryophytes. The gametophyte of angiosperms depends totally on the sporophyte. Angiosperms are important to humans because world economy is overwhelmingly based on them.

Botanists commonly divide angiosperms into monocots and dicots. The typical features of these groups of plants are described in table 30.1.

Having studied gymnosperms, you might conclude that the vegetative adaptations of angiosperms and gymnosperms are similar. This is true. However, in this exercise you will study two unique adaptations of angiosperm reproduction: flowers and fruit.

STRUCTURE AND FUNCTION OF FLOWERS

Angiosperms have a seemingly infinite variety of flowers, ranging from the microscopic flowers of *Lemna* (duckweed) to the rare, monstrous blossoms of *Rafflesia*, which measure up to 1 m across. In this exercise, we'll consider only the "typical" flower depicted in figure 30.1. Examine the flower model(s) available in the lab and the living flowers on display, and identify the following parts:

-   **peduncle**—flower stalk

-   **receptacle**—the part of the flower stalk that bears the floral organs; located at the base of the flower; usually not large or noticeable

-   **sepals**—the lowermost or outermost structures, which are usually leaflike and protect the developing flower; the sepals collectively constitute the **calyx**

-   **petals**—located inside and usually above the sepals; may be large and pigmented (in insect-pollinated flowers) or absent (in wind-pollinated plants); the petals collectively constitute the **corolla**

TABLE 30.1

CHARACTERISTICS OF THE TWO CLASSES OF ANGIOSPERMS

Class Liliopsida (monocots)	Class Magnoliopsida (dicots)
One cotyledon per embryo	Two cotyledons per embryo
Flower parts in sets of three	Flower parts in sets of four or five
Parallel venation in leaves	Reticulate (netted) venation in leaves
Multiple rings of vascular bundles in stem	One ring of vascular bundles or cylinder of vascular tissue in stem
Lack of a true vascular cambium	Possess a true vascular cambium

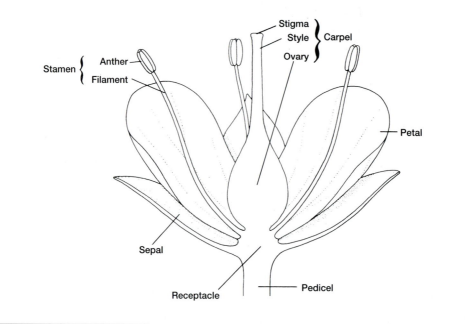

FIGURE 30.1

"Typical" flower.

- **perianth**—the combined calyx and corolla
- **androecium**—the male portion of the plant, which arises above and inside the petals; consists of **stamens,** each of which has a **filament** topped by an **anther;** inside the anthers are **pollen grains** which produce the male gametes
- **gynoecium**—the female portion of the plant, which arises inside and above the androecium; consists of one or more **carpels,** each made up of an **ovary, style,** and **stigma;** the ovary contains **ovules,** which contain the female gametes; during pollination, pollen grains are transferred to the stigma, where they germinate and grow a tube through the style to the ovary

The sepals and petals are usually the most conspicuous parts of a flower, and a variety of flower types are described by the characteristics of the perianth. In **regular** flowers, all parts of the perianth are similar in size and shape and are **radially symmetrical,** or arranged like a wheel. In **irregular** flowers, some of the perianth parts are a different size and shape. An irregular flower is **bilaterally symmetrical** because the flower has obvious right and left sides. If the petals, sepals, and stamens originate below the ovary, the flower is **hypogynous.** If they originate above the ovary, the flower is **epigynous.**

Procedure 30.1
Examine flowers and their parts

1. Obtain a flower provided in the laboratory.
2. Remove the parts of the flower, beginning at the lower outside.
3. Locate the petals and sepals, and determine their arrangement and point of attachment.

Question 1
a. Is the flower regular or irregular? _____

b. Is it radially or bilaterally symmetrical? _____

30-2

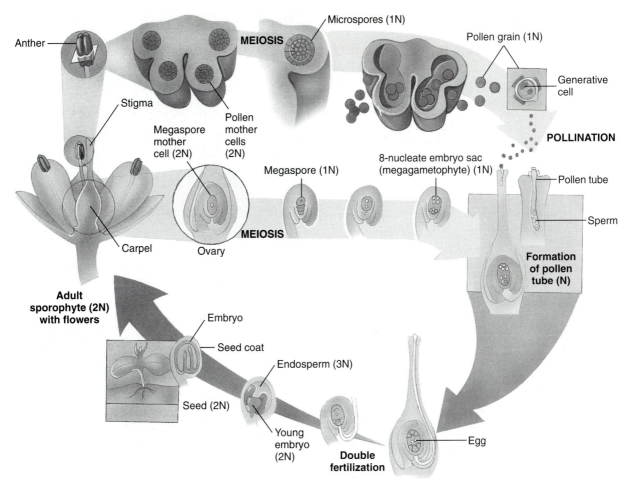

Labels in figure:
Anther — MEIOSIS — Microspores (1N) — Pollen grain (1N) — Generative cell — POLLINATION — Pollen tube — Sperm — Stigma — Megaspore mother cell (2N) — Pollen mother cells (2N) — Megaspore (1N) — 8-nucleate embryo sac (megagametophyte) (1N) — MEIOSIS — Carpel — Ovary — Formation of pollen tube (N) — Adult sporophyte (2N) with flowers — Embryo — Seed coat — Endosperm (3N) — Seed (2N) — Young embryo (2N) — Double fertilization — Egg

FIGURE 30.2

Angiosperm life cycle. As in pine, the sporophyte is the dominant generation. Eggs form within the megagametophyte, or embryo sac, inside the ovules, which in turn are enclosed in the carpels. The carpel is differentiated into a slender style, ending in a stigma. Pollen grains stick to the surface of the stigma and germinate. Pollen grains form within the sporangia of the anthers and complete their differentiation either before or after grains are shed. Fertilization is a double process. A sperm and an egg fuse to produce a diploid (2N) zygote; at the same time, another sperm fuses with the two polar nuclei, producing the primary endosperm nucleus, which is triploid (3N). Both the zygote and the primary endosperm nucleus divide mitotically, giving rise, respectively, to the embryo and the endosperm. The endosperm is unique to angiosperms and nourishes the embryo and young plant.

c. Is it hypogynous or epigynous? _____

4. Remove the petals and sepals.

5. Locate and remove a stamen and place it on a slide. Examine this dry mount of the stamen on a microscope using low magnification.

6. Add a drop of water to the preparation, and open the anther if necessary to disperse the pollen. Cover with a coverslip.

7. Use high magnification to locate the generative and tube nucleus of a pollen grain. The **generative nucleus** is usually small, spindle-shaped, and off center. The **tube nucleus** is larger and centered. You'll learn more about pollen nuclei later in this exercise.

8. To enhance visibility of the nuclei, stain the preparation with acetocarmine. Add 1 or 2 drops to the preparation and gently heat it. Do not overheat. Reexamine the preparation.

9. Locate the gynoecium of the flower, and make longitudinal and transverse sections. The gynoecium consists of one or more fused carpels, each having an interior cavity called a **locule,** which contains ovules.

Question 2

a. How many carpels (locules) are apparent? _____

b. How many ovules are developing in each locule?

_____

10. Examine the other types of flowers available in the lab. Repeat the previous steps to guide your examination.

Alternation of Generations in Flowering Plants

The life cycle of flowering plants involves the alternation of a multicellular haploid stage with a multicellular diploid stage, as is typical for all plants (figs. 29.1 and 30.2). The

30-3

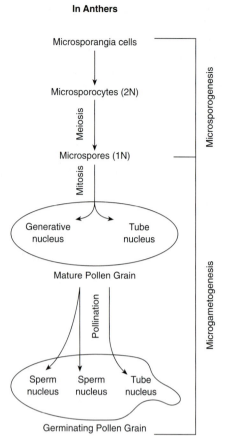

In Anthers

Microsporangia cells

↓

Microsporocytes (2N)

Meiosis

Microspores (1N)

Mitosis

Generative nucleus — Tube nucleus

Mature Pollen Grain

Pollination

Sperm nucleus — Sperm nucleus — Tube nucleus

Germinating Pollen Grain

Microsporogenesis

Microgametogenesis

FIGURE 30.3

Microsporogenesis and microgametogenesis in the anthers of flowers. Mature pollen grains (see figure 30.4) are microgametophytes. See figure 30.5 for micrographs of microsporogenesis and microgametogenesis.

diploid **sporophyte** produces haploid spores by meiosis. Each haploid spore grows into the **gametophyte,** which produces gametes by mitosis and cellular differentiation. In angiosperms the sporophyte is the large, mature organism with flowers that you easily recognize. The gametophyte within the flower is reduced to a pollen grain (which produces a sperm nucleus) or an ovule (which produces an egg). Review your textbook's description of alternation of generations.

Production of spores in the sporophyte by meiosis is part of a larger process called **sporogenesis.** Flowering plants produce two kinds of spores: **microspores** and **megaspores.** Production of gametes by the gametophytes is called **gametogenesis.**

Microsporogenesis, Production of Pollen, and Microgametogenesis

Microsporogenesis is the production of microspores within **microsporangia** of a flower's anthers via meiosis of **microsporocytes,** or microspore mother cells (fig. 30.3). These microspores grow and mature into microgametophytes, also known as pollen grains (fig. 30.4). The haploid nuclei in a

FIGURE 30.4

Pollen grains of ragweed are spherical and elaborately sculptured. Their outer wall contains proteins that regulate germination of the pollen grain when it lands on a stigma of a ragweed flower. These same proteins also cause allergies in many people.

mature pollen grain include a tube nucleus (or vegetative nucleus) and a generative nucleus. The pollen grain germinates when it lands on a flower stigma, and the tube nucleus controls the growth of the pollen tube. The generative nucleus replicates to produce two **sperm nuclei.**

Pollen causes allergies in many people. However, studies of pollen are important in science beyond the treatment of allergies. For example, geologists examine the color of pollen brought up in sediment cores during oil drilling. Dark brown to black pollen indicates temperatures too high for oil deposits and that the well will likely produce natural gas. Orange pollen indicates less intense heat, which is associated with high-quality oil production. In addition, examination of fossil pollen tells us about the diversity of ancient floras and helps us locate ancient seas and their shorelines where pollen is known to accumulate.

Procedure 30.2
Examine stages of microsporogenesis and microgametogenesis in *Lilium*

1. Examine fresh or preserved specimens of dehiscent and predehiscent anthers.

Question 3

a. Based on your observations, how would you define dehiscent and predehiscent? _____

_____

b. What differences can you detect between the structures of dehiscent and predehiscent anthers?

_____

_____

c. Which stage is the most mature? _____

(a)

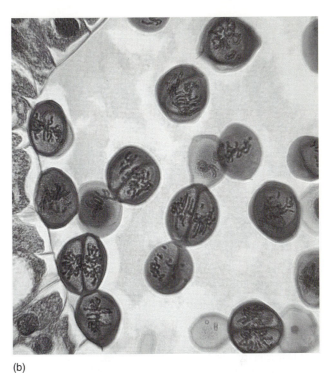

(b)

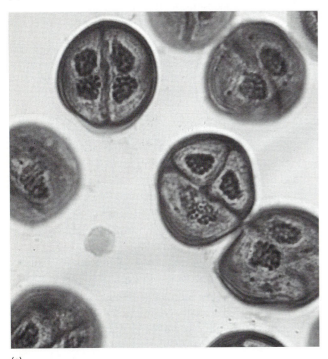

(c)

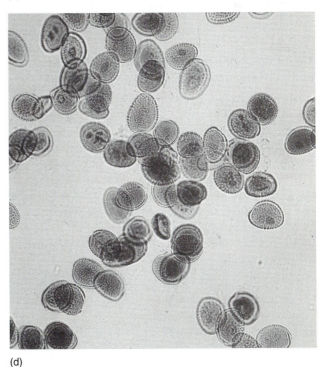

(d)

FIGURE 30.5

Stages of microsporogenesis and microgametogenesis in *Lilium*. (*a*) Cells in early prophase I. (*b*) Cells in second meiotic division. (*c*) Tetrads of microspores from meiosis. (*d*) Mature pollen. For an overview of these stages see figure 30.3.

2. Examine a prepared cross section of a young anther. Note the immature sporangia tissue, which will form microsporocytes.

3. Examine a prepared cross section of a lily anther showing microsporocytes in early prophase I (fig. 30.5*a*).

4. Examine a prepared slide showing stages of meiosis II (fig. 30.5*b*).

5. Examine a prepared slide showing pollen tetrads of microspores produced by meiosis (fig. 30.5*c*).

6. Examine a prepared slide showing mature pollen with two or more nuclei (fig. 30.5*d*).

7. Examine living or prepared pollen from various plants if available. Note any differences among pollen grains and differences between pollen of monocots and dicots.

Procedure 30.3
Observe germination of pollen grains

1. At the beginning of the lab period place some pollen in a drop of 1% sucrose within a ring of petroleum jelly on a microscope slide. Alternatively, a preparation of pollen germinating on 5% sugar and agar may be available for observation.

2. Apply a coverslip and examine the pollen intermittently with your microscope throughout the lab period. Incubate the slide in a warm place.

3. Ring the coverslip with petroleum jelly to prevent the preparation from drying.

Question 4
a. How many pollen grains germinated? _____

b. Can you see vegetative and generative nuclei in the pollen tubes? _____

Megasporogenesis, Production of Ovules, and Microgametogenesis

Megasporogenesis is the production of megaspores (fig. 30.6). It occurs in the sporangia of the flower ovary by meiosis of **megasporocytes** (megaspore mother cells). These megaspores undergo **megagametogenesis;** that is, they develop into **megagametophytes,** which produce egg gametes. The megagametophyte and its surrounding tissues is called an **ovule.** Ovules usually have two coverings called **integuments.** The entire haploid structure is called the **embryo sac** and will consist of only six to ten nuclei, one of which is an egg. A seven- or eight-celled embryo sac is most common (fig. 30.7).

Procedure 30.4
Examine stages of megasporogenesis and megagametogenesis in *Lilium*

1. Examine a prepared cross section of a *Lilium* ovary showing a diploid megasporocyte within the sporangium before meiosis (fig. 30.8a).

2. Examine a prepared slide showing the four-nucleate embryo sac after meiosis (fig. 30.8b). Meiosis produces four haploid megaspores in the embryo sac.

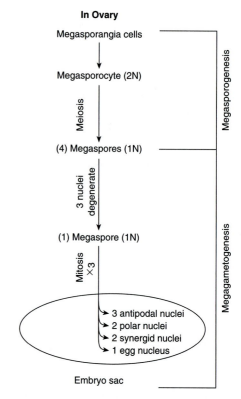

FIGURE 30.6

Megasporogenesis and megagametogenesis in the ovaries of flowers. The embryo sac is the mature megagametophyte. The locations of the processes are shown in figures 30.7 and 30.8.

In *Lilium* these four nuclei will undergo mitosis, but this is not typical. In most other angiosperms, three of the four nuclei degenerate, and the single remaining nucleus passes through two mitotic divisions before the next stage.

3. Examine a prepared slide showing the eight-nucleate embryo sac (fig. 30.8c). Because of the large size of the embryo sac, it is seldom possible to observe all eight nuclei in the same section.

4. At one end of the megagametophyte toward the micropyle (the small opening where the pollen tube enters the ovule), locate the eight nuclei including the **egg** and two **synergid nuclei** associated with fertilization. At the opposite end, locate three **antipodal cells** that ordinarily do not participate in reproduction. In the center are two **polar nuclei,** which migrated from each pole of the megagametophyte.

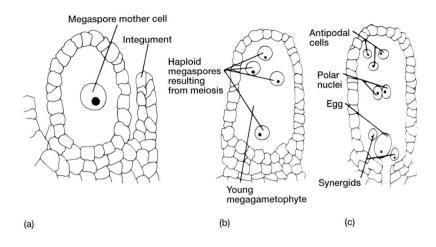

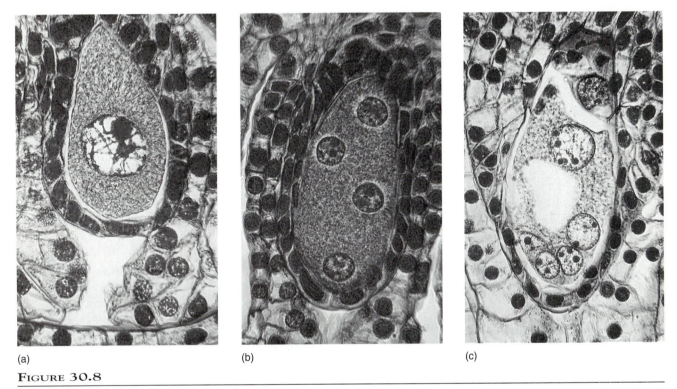

FIGURE 30.7

Megasporogenesis and megagametogenesis in *Lilium*. (*a*) The megaspore mother cell (megasporocyte) is diploid and undergoes meiosis. (*b*) Four haploid megaspores result from meiosis. (*c*) The mature megagametophyte contains eight nuclei, one of which is the egg.

(a)   (b)   (c)

FIGURE 30.8

Stages of megasporogenesis and megagametogenesis in *Lilium*. (*a*) Diploid megasporocyte in a lily ovary. (*b*) Embryo sac with four nuclei produced by meiosis. (*c*) Embryo sac (mature megagametophyte) with eight nuclei. For an overview of these stages see figure 30.6.

POLLINATION AND FERTILIZATION

Botanists have long been interested in plant pollination for obvious economic reasons. Angiosperm reproduction depends on pollination. After the development of a microgametophyte (pollen grain) with sperm and a megagametophyte (ovule) with egg, sexual reproduction in angiosperms typically occurs as follows:

1. Pollination occurs when pollen is transported to the surface of the flower's stigma.

2. The pollen grain germinates, and a pollen tube grows through the stigma and style to the ovary. Its growth is governed by the style and the tube nucleus of the pollen grain.

3. One sperm nucleus fuses with the egg to form the diploid zygote, and the other sperm nucleus fuses with the two polar nuclei. This process, called **double fertilization,** is characteristic of angiosperms.

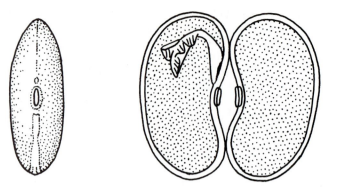

FIGURE 30.9

Bean seed.

4. The zygote develops into the embryo. The fusion of sperm nucleus and two polar nuclei forms the triploid endosperm that provides food for the embryo.

5. The integuments of the ovule form the seed coat, and the fruit develops from the ovary and other parts of the flower.

SEED STRUCTURE

Procedure 30.5
Observe parts of a bean seed

1. Obtain some beans that have been soaked in water for 24 hours.

2. Peel off the seed coat, and separate the two cotyledons. Between the cotyledons you'll see the young root and shoot.

3. Label the following features on figure 30.9:

   micropyle—a small opening on the surface of the seed through which the pollen tube grew

   hilum—an adjacent eliptical area at which the ovule was attached to the ovary

   cotyledon—food for the embryo

   embryo with young root and shoot—develops into the new sporophyte

4. Add a drop of iodine to the cut surface and observe the staining pattern. Indicate this pattern on figure 30.9.

5. Repeat steps 2 to 4 with soaked peas.

Question 5

a. How are seeds of peas and beans similar? _____

_____

How are they different? _____

_____

b. Based on the appearance of seeds that you have examined, how do you think each is dispersed (by insects, wind, etc.)? _____

_____

FRUIT

A **fruit** is a mature, ripened ovary plus any associated tissues. Therefore, a fruit contains seeds. Apples, peaches, tomatoes, and okra are all examples of fruit. A mature fruit is often larger than the ovary at the time of pollination and fertilization, which indicates that a great deal of development occurs while the seeds are maturing. Most fruits are either **dry** or **fleshy**. Dry fruits crack or split at maturity and release their seeds. Sometimes the dry wall surrounds the seed until it germinates. The fruit wall is usually tough and hard and is sometimes referred to as stony. The seeds of fleshy fruits remain in the tissue until germination.

A typical fruit has an outer wall called a **pericarp,** which is composed of an **exocarp, mesocarp,** and **endocarp.** Within the pericarp are seeds, various partitions, and **placental tissues.** The fruit often includes the receptacle of the flower.

Procedure 30.6
Observe diversity of fruits

1. Use the following descriptions to study and classify fruits available in the lab.

   I. Simple fruit—derived from one carpel or several united carpels

   A. Dry fruit—fruit wall becomes papery or leathery at maturity

   1. Dehiscent—fruit has seams that open at maturity

a. Follicle—derived from one carpel; splits open along one seam

   Examples: milkweed, larkspur

b. Legume—derived from one carpel; splits open along two seams

   Examples: peanut, mesquite, pea pod

c. Capsule—derived from more than one carpel; opens by various means

   Examples: castor bean, cotton

2. Indehiscent—fruit doesn't open along seams at maturity

   a. Grain or caryopsis—seed coat fused with ovary wall

      Example: corn

   b. Nut—ovary wall is woody

      Examples: acorn, pecan

   c. Achene—seed coat not fused with ovary wall

      Example: sunflower

   d. Samara—ovary wall forms a winglike structure

      Examples: ash, maple

B. Fleshy fruit—ovary wall is fleshy at maturity

   1. Superior ovary—ovary is free and separated from the calyx

      a. Berry—ovary wall is fleshy

         Examples: avocado, tomato, grape

      b. Hesperidium (false berry)—berry with leathery separable rind

         Examples: citrus fruit

      c. Drupe—same as berry but with woody interior coating

         Examples: peach, plum, olive

   2. Inferior ovary—ovary is more or less united with the calyx

      a. Pepo (false berry)—ovary wall and floral tube become fleshy

         Examples: cucumber, squash, banana

      b. Pome—same as false berry but with leathery interior coating

         Examples: apple, pear

II. Aggregate fruit—derived from a flower with several carpels; ovaries unite to form a single structure

   Examples: strawberry, raspberry

III. Multiple fruit—derived from a cluster of flowers

   Examples: fig, pineapple

Procedure 30.7

Observe the structure of a bean, sunflower, corn grain, apple, and tomato

1. Examine the pod of a stringbean. Notice that this single carpel has two seams that can open to release seeds. Remove the seeds; then locate and remove the seed coat. Split the seed and locate the embryo.

Question 6

a. Does the pod appear to be a single carpel with one locule containing seeds? _____

b. Is the micropyle near the attachment of the seed to the pod? _____

2. Crack the outer coat of a sunflower fruit and remove the seed. Locate the embryo and determine whether sunflower is a monocot or dicot.

Question 7

a. Before today would you have referred to the uncracked sunflower achene as a fruit? _____

   Why or why not? _____

   _____

b. Is a sunflower a monocot or dicot? _____

3. Examine an ear of corn from which the husks have been removed without disturbing the "silk." The strands of silk are styles of the gynoecium. Examine a corn grain. Section the grain longitudinally. Identify the endosperm and embryo. Refer to exercise 19 for more information on the structure of the grain.

Question 8

a. If a corn grain is actually a fruit, where is the pericarp? _____

b. Does a corn grain have a cotyledon as well as an endosperm? _____

4. Examine an apple, which is an example of a pome. Section the apple longitudinally and transversely. Compare its structure with figure 30.10. Most of the outer flesh of the apple is derived from tissue other than the ovary. Locate the outer limit of the pericarp and the limit of the endocarp.

Question 9

a. Is a potato derived from an ovary? _____

b. How many carpels are fused to form an apple? _____

c. How might the fleshy pericarp aid in seed dispersal?

   _____

   _____

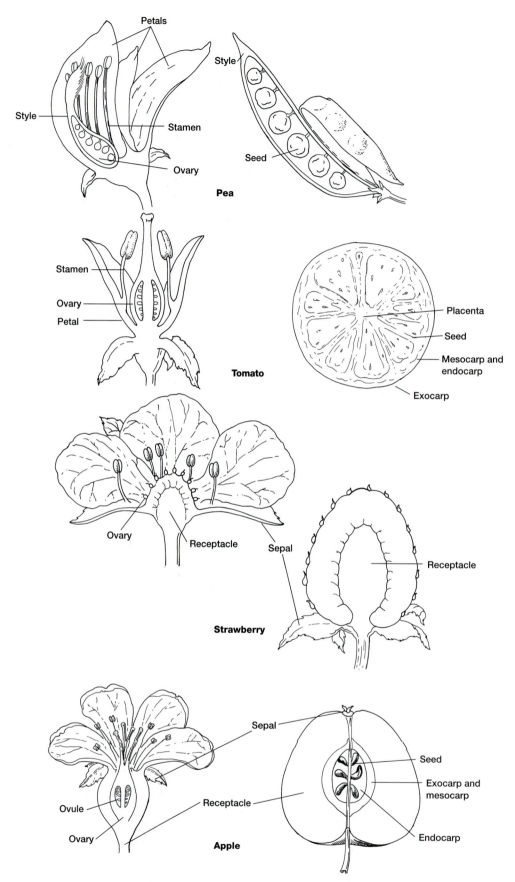

Figure 30.10

Diagrams of fruit and their originating flowers.

5. Examine a tomato, and section it transversely. Compare its structure with that shown in figure 30.10. The jellylike material is the placenta giving rise to the ovules.

Question 10

a. How many carpels are fused to form a tomato?

_____

Summary of Activities

1. Examine models of flower parts.
2. Examine wet mounts and prepared slides of anther cross sections.
3. Examine living and prepared pollen from various plants.
4. Observe germinating pollen grains.
5. Examine prepared slides of stages of developing ovules.
6. Stain and examine parts of bean seeds.
7. Examine and classify diverse fruits according to their structure.

Questions for Further Thought and Study

1. What is meant by "double fertilization"?
2. What features of seeds and fruits have allowed angiosperms to become so widespread?
3. What are the similarities and differences between cones and flowers?
4. What is the difference between a fruit and a vegetable?
5. Diagram the life cycle of an angiosperm. Which parts are haploid? Which are diploid?

31

Pollution

Objectives

By the end of this exercise you will be able to

1. describe how acid rain, chemical pollution, and thermal pollution affect the growth of organisms

2. determine whether a water sample is polluted by algae

3. assay water samples for potential toxicity to plant growth.

Seldom a day passes in which we don't hear a pollution-related story in the news. Oil spills, leaks of toxic chemicals, and noise pollution are just a few of the pollution-related problems that affect our lives. The effects of pollution can be disastrous, as exemplified by the nuclear explosion at Chernobyl, the Alaskan oil spills, and the dumping of toxic wastes at the Love Canal. Whether or not it makes headlines, pollution of any kind affects fundamental aspects of life, namely how organisms grow and interact. Because populations of organisms interact among themselves and with their environment, pollution always affects more than one organism. A pollutant that reduces a certain population, especially plants at the base of the food chain, significantly affects all other levels of the food chain. Thus, water polluted with chemicals such as herbicides, which kill algae, decreases populations of zooplankton, which feed on the algae; this ultimately affects populations of fish and humans, which are part of the food chain.

A **pollutant** is any physical or chemical agent that decreases the aesthetic value, economic productivity, or health of the biosphere. There are many kinds of pollutants, such as noise, chemicals, radiation, and heat. In this exercise, you will study several of these types of pollution. Specifically, you will

- simulate the effects of acid precipitation by examining seed germination at acidic pH;

- study chemical pollution by examining the effects of differing concentrations of nutrients and pesticides on growth of algae;

- examine water polluted with algae;

- use the *Allium* test to assay a variety of pollutants.

SIMULATING THE EFFECTS OF ACID RAIN

Acid rain is a worldwide problem caused by atmospheric pollution. Compounds such as nitrates and sulfates released into the atmosphere by automobiles and industries combine with rainwater to form nitric and sulfuric acids that are deposited across the countryside in precipitation. Acid rain decreases the pH of the soil, affects the availability of nutrients, and typically diminishes plant growth (see plate II, fig. 3). The effects of acid rain are often subtle, but they are significant; the cumulative pollution of decades of acid rain are now killing forests and reducing crop yields in many areas.

In this exercise, you will simulate acid rain with solutions of sulfuric acid and investigate the influence of acid rain on seed germination.

CAUTION

Be careful handling the cultures; do not get acid solutions on yourself. If you do, wash thoroughly and immediately.

TABLE 31.1

GERMINATION OF SEEDS IN ENVIRONMENTS OF DIFFERENT pH

pH	Total Number of Seeds	Number Germinated	% Germination
7	____	____	____
4	____	____	____
2	____	____	____

TABLE 31.2

EFFECTS OF NUTRIENT ENRICHMENT ON ALGAL CULTURES

Culture	Observations
1 (nutrient-enriched)	
2 (nutrient-enriched, anaerobic)	
3 (control)	

Procedure 31.1
Observe seed germination

1. Examine the three petri plates labeled "Acid Rain and Seed Germination." All of the dishes were inoculated several days ago with fifty seeds of corn (*Zea mays*). Dish 1 contains seeds soaked in a solution of pH 4, and dish 2 contains seeds soaked in a solution of pH 2.

2. Compare these seeds with those grown in dish 3 at neutral pH (pH 7).

3. Use the following formula to calculate the percentage germination for each treatment:

$$\% \text{ germination} = (\text{seed}_{\text{germ}}/\text{seeds}_{\text{tot}}) \times 100$$

where,

$\text{seeds}_{\text{germ}}$ = number of germinated seeds

$\text{seeds}_{\text{tot}}$ = total number of seeds

4. Record your results in table 31.1.

Question 1
a. How does acidity affect seed germination? _____

_____

b. What does this tell you about the effect of acid rain on seed germination in natural environments?

_____

_____

c. How would the decreased growth of plants in response to acid rain affect animals in the same environment? _____

_____

NUTRIENT ENRICHMENT: EUTROPHICATION

Bodies of water such as lakes and ponds naturally undergo a nutrient-enrichment process called **eutrophication.** In extreme situations, this enrichment stimulates algal and plant growth so much that the lake or pond may eventually fill in and disappear. Human-generated pollutants such as raw sewage and fertilizers dramatically speed eutrophication. In this exercise, you'll study eutrophication by studying the growth of organisms in nutrient-enriched media.

Several days ago algal cultures were started with similar amounts of algae. Culture 1 contains an excess of nitrogen, a condition similar to that of lakes enriched with raw sewage. Culture 2 contains large amounts of nitrogen and minimal oxygen to simulate anaerobic conditions that occur naturally when bacterial populations in water increase rapidly in response to added sewage. As the bacteria grow and reproduce, they rapidly deplete the oxygen supply in the water, thus producing anaerobic conditions. Culture 3 is an aerobically grown control containing normal amounts of nitrogen.

Procedure 31.2
Examine the effects of simulated eutrophication

1. Examine the two algal cultures labeled "Eutrophication" and "Control."

2. Carefully examine each of the cultures and record your observations in table 31.2.

Question 2

a.   Which culture has the most algal growth? _____

b.   Which has the least? _____

c.   What does this tell you about the influence of

   eutrophication on algal growth? _____

   _____

d.   How could you reverse the effects of eutrophication?

   _____

e.   Can eutrophication ever be beneficial? _____

   If so, when? _____

   _____

ALGAL POLLUTION OF WATER SUPPLIES

Increasing human populations have created greater demands on groundwater supplies. To circumvent these demands, many areas have started using surface waters such as ponds and lakes to meet their water demands. Unlike groundwater, which is usually free of algae, surface water contains large amounts of algae, which produce tastes and odors (*Anabaena, Nitella, Pandorina, Hydrodictyon*), clog pipes (*Synedra, Anacystis*), produce toxins (*Anabaena*), and form mats covering the lake or reservoir (*Oscillatoria, Spirogyra*). Work in small groups to assess algal pollution in the water samples that are available in lab.

Procedure 31.3

Determine algal pollution of water samples

1.   Dip a filter base and funnel (fig. 31.1) in 70% ethanol for 1 minute. Shake off the excess alcohol and let the apparatus air-dry.

2.   Use sterilized forceps to place a filter with pores 0.45 μm in diameter on the filter base. Screw the funnel onto the support (fig. 31.1) so that the filter is held securely in place.

3.   Add 200 mL of the water sample to the funnel. If the water sample is heavily polluted, decrease the sample to 100, 50, or 25 mL. Use a hand vacuum pump or water aspirator (fig. 31.2) to evacuate the receiving flask. Algae will be trapped on the filter.

4.   Remove and dry the filter in an oven at 45°C for 8 minutes.

5.   Float the dried filter on approximately 5 mL of immersion oil in a small petri plate. The dry filter will become transparent in the oil. If the filter remains opaque, place the petri plate on a warm surface until it clears.

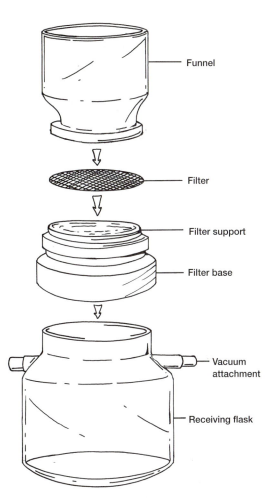

FIGURE 31.1

Diagram of filtration apparatus used to collect organisms from a water sample.

6.   Remove the filter from the petri plate by dragging it across the side of the plate.

7.   Cut the filter into fourths, and place these pieces of filter all on one clean microscope slide.

8.   Examine the filter with reflected light at low total magnification (10–40×).

9.   Count the number of algae in each of ten randomly selected fields of view and record the numbers in the following space:

Number of Algae in Each Field of View

Field	1	2	3	4	5	6	7	8	9	10
Number of algae	—	—	—	—	—	—	—	—	—	—

Total ___

10.   Calculate the area of the field of view with a stage micrometer or ruler (see exercise 1):

Area of field of view = ___

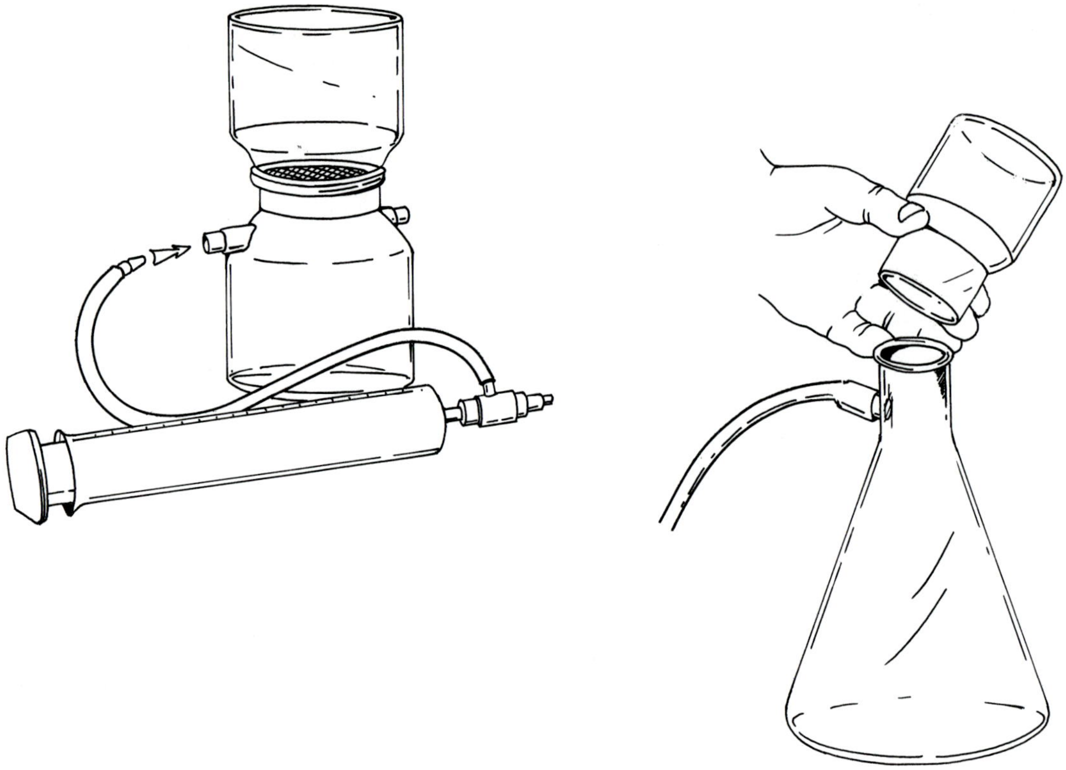

(a)   (b)

FIGURE 31.2

Vacuum filtration using (*a*) a hand vacuum pump and (*b*) a water aspirator.

11. Use the following formula to calculate the number of algae on your filter. The value 1,380 mm^2 is the area of the 47 mm filter that you used to collect the algae.

$$\frac{\text{Total no. of}}{\text{algae on filter}} = \frac{1,380 \text{ mm}^2}{\text{Area of one field of view (mm}^2)} \times \frac{\text{Total no. of algae counted}}{\text{No. of fields counted}}$$

Total no. of algae on filter = _____

12. Use the following formula to calculate the number of algae per milliliter of sample:

$$\text{No. of algae mL}^{-1} = \frac{\text{Total no. of algae on filter}}{\text{Volume of sample (mL)}}$$

No. of algae mL^{-1} = _____

Algal populations exceeding 1,000 organisms per mL indicate that the water is overenriched by sewage or other nutrients.

Question 3

a. How many algae per milliliter are in your water sample? _____

b. Is your sample polluted? _____

c. In what ways might these algae benefit other organisms in the water? _____

_____

d. In what ways might they harm other organisms?

_____

_____

e. Suppose you are responsible for "cleaning up" a lake that is polluted with algae. What would you do, and

why? _____

_____

POLLUTANTS AS GROWTH INHIBITORS

Many pollutants affect the growth and development of organisms. Thus, the most informative assay involves living organisms growing under controlled treatments of suspected pollutants. The *Allium* test provides this kind of assay. In the *Allium* test, bulbs of the common onion (*Allium*) are subjected to solutions of water being investigated for pollutant toxicity. Roots develop quickly, and their number and length after 5 days measures the effect of the potential pollutant on a common aspect of plant growth.

TABLE 31.3

ROOT GROWTH OF REPLICATE ALLIUM BULBS EXPOSED TO A SPECIFIC TREATMENT SOLUTION

Treatment: _____   Control: _____

Number of Roots	Length of Roots	Number of Roots	Length of Roots
___	___	___	___
___	___	___	___
___	___	___	___
___	___	___	___
___	___	___	___
___	___	___	___
___	___	___	___
___	___	___	___
Mean = ____	Mean = ____	Mean = ____	Mean = ____

Pesticides

Pesticides are common pollutants; however, the increasing demands for food by our expanding population has caused farmers to rely extensively on pesticides to increase crop yields. These pesticides are often effective, but usually they are washed from the soil by rain and contaminate lakes, ponds, and the underground water supply. These pollutants harm organisms that live in these waters.

Pesticides are only a small percentage of the kinds of chemicals introduced into the environment. Consider that over 60,000 synthetic chemicals are currently in use, and over 1,000 new chemicals are being synthesized every year. The potential harm from these chemicals requires that we carefully assess their effects on living processes.

CAUTION

Do not open the pesticide cultures or spill the liquids on yourself. Pesticides are somewhat toxic.

Thermal Pollution

Excessive heat is a common pollutant. Many factories use water to cool heat-generating equipment. Water used for this cooling is then released into reservoirs or ponds, where it raises the water temperature. Thermal pollution

- speeds biochemical reactions, thereby altering the growth and composition of populations;
- speeds evaporation of water, thereby concentrating other pollutants in the water;
- decreases the oxygen supply in the water, because warmer water holds less oxygen than does cooler water.

Procedure 31.4
Use the *Allium* test to assay a variety of pollutants

1. Examine the solutions provided by your instructor for you to assay with the *Allium* test. Record the solutions in table 31.3.

2. You may need to work in small groups to assay all of the solutions. Determine which solutions or treatments will be each group's responsibility. Some of the treatments may involve different temperatures and some may be solutions with unknown contents.

3. Obtain onion bulbs. You will need five to ten bulbs for each treatment and five to ten bulbs for controls for each treatment.

4. Trim the outer, loose layers of each bulb. Use a razor blade to trim exposed tissue from the root crown. Your instructor will demonstrate how to do this.

5. Obtain a beaker (or test tube) that is wide enough to support the onion but will allow the root crown to protrude into the solution.

6. Fill five to ten replicate beakers with the solution being tested.

7. Fill five to ten beakers with the appropriate control solutions.

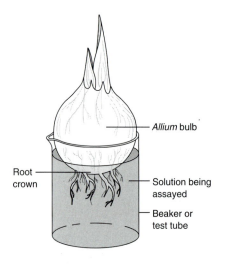

Root crown

Allium bulb

Solution being assayed

Beaker or test tube

FIGURE 31.3

Allium test setup. The number and length of the growing roots quantifies the toxicity of the solution being assayed.

TABLE 31.4

ROOT GROWTH OF ALLIUM BULBS EXPOSED TO SOLUTIONS CONTAINING DIFFERENT POLLUTANTS

Solutions Assayed	Mean Number of Roots		Mean Length of Roots	
	Treatment	Control	Treatment	Control
_____	_____	_____	_____	_____
_____	_____	_____	_____	_____
_____	_____	_____	_____	_____
_____	_____	_____	_____	_____
_____	_____	_____	_____	_____
_____	_____	_____	_____	_____

Question 4

a. What is the appropriate control solution for your treatment? _____

8. Put a trimmed onion bulb on your beaker so the root crown is completely submerged (fig. 31.3).

9. Incubate the treatments and controls in the dark at room temperature for 3 to 5 days.

10. Check and supplement the solution levels in the beakers periodically if needed.

11. After 3 to 5 days, determine the mean root length and number of roots for each bulb. Record the values in table 31.3. These are the results for the treatment solution that your group tested.

12. Record your mean values from table 31.3 in table 31.4.

13. In table 31.4 record the results for the solutions tested by the other groups.

14. Compare the number and length of roots on treated bulbs with those of controls.

Question 5

a. Why did each group need to run controls, instead of using one set of controls for everyone? _____

b. The *Allium* test is a common test. What are some disadvantages to this test? _____

_____

c. The *Allium* test for thermal pollution may show that higher temperatures accelerate growth. Is this bad?

_____

Why or why not? _____

_____

Summary of Activities

1. Observe the effects of pH on seed germination.
2. Observe the effect of excess nutrients on algal cultures.
3. Measure the level of algal pollution in water samples.
4. Use the *Allium* test to assay water samples for toxicity.

Questions for Further Thought and Study

1. What causes pollution?
2. Is all pollution bad? Why or why not?
3. Polluting lakes with laundry water and sewage often produces an algal "bloom." Why don't populations of predators (i.e., zooplankton and fish) increase to offset this bloom and keep the algal population in check?

Doing Botany Yourself

Design your own assay similar in concept to the *Allium* test. Use a living plant and test potential pollutants of your choice.

Writing to Learn Botany

Define pollution. Why would a material be a pollutant in some situations but not in others?

32

Community Ecology

Objectives

By the end of this exercise you will be able to

1. observe the physical factors, plant dominance, and interactions among organisms in a terrestrial community and characterize the community according to these factors;

2. quantify the distribution and abundance of plants in a community;

3. detect the experimental effects of competition and allelopathy among plants.

Ecological communities are extraordinarily complex. The assemblage of plants that you observe at any given time results from interactions among plants and their physical surroundings, among different species of plants, and among plants and animals. All of these interactions are driven by a flow of energy captured by green plants and passed to herbivores, predators, and decomposers.

It is beyond the scope of this exercise to explain and categorize all the processes occurring in a plant community, but we can assemble some basic observations that characterize and distinguish communities.

QUALITATIVE COMMUNITY ASSESSMENT

Procedure 32.1
Observe and assess the ecological characteristics of a terrestrial community

1. Locate and visit a terrestrial community designated by your instructor.

2. Characterize the community according to the criteria and questions that follow. After you've answered the questions, discuss your observations with your instructor and other groups. Be prepared to describe your assessment of energy flow through the community, diversity of the community, and interactions among organisms in the community.

Physical Factors
How would you assess the light intensity? _____

_____

Does the community include shade-tolerant as well

as shade-intolerant plants? _____

How might apparent shade affect the temperature of

the community? _____

_____

How does light differ in different vertical layers of

vegetation? _____

Why could this be important? _____

_____

What is the temperature 2 meters aboveground?

_____

What is the temperature at the soil surface? _____

What parts of the community might be cooler than

others? _____

Why could this be important? _____

_____

How does the ground slope? _____

Why could this be important? _____

_____

How would you characterize the soil (e.g., loam, clay, sand)? _____

Based on your observations of slope and soil type, would you expect the soil to retain moisture? _____
What is the nature of the ground cover (e.g., grasses, bare soil)? _____
_____

Is there a layer of leaf litter on the ground? _____
Is the community generally a moist, moderate, or dry environment? _____

Plant Dominance

Which plant species are most abundant in terms of numbers? _____
Which plant species are most abundant in terms of biomass? _____

Would you describe this community as diverse? _____
Why or why not? _____
_____

What general categories of plant types (e.g., shrubs, trees) are apparent? _____
What is the vertical distribution of vegetation?
_____

Interactions Among Organisms

What evidence do you see of resident vertebrates?
_____

What evidence do you see of resident invertebrates?
_____

If you don't see any vertebrates, does that mean there are none? _____
What evidence do you see of plant–animal interactions? _____
What evidence do you see of plant–plant interactions? _____
What adaptations do the plants have to discourage herbivores? _____
_____

Do you see any obvious or subtle evidence of competition by plants for available resources? _____
_____

Reexamine your observations just listed. How would each affect the type and growth of plants in the community that you studied? _____
_____

QUANTITATIVE COMMUNITY ASSESSMENT

A variety of techniques have been developed to measure the numbers, densities, and distributions of organisms in terrestrial plant communities. One widely used technique is to count organisms within randomly distributed **quadrats** (sometimes called plots) of uniform size. Other techniques such as the **line-intercept method** involve measuring distances between plants or the distance from randomly chosen points to the nearest plants. In the line-intercept method a **transect,** or line, is established and laid out within the community. Organisms that touch this line are counted and measured. Calculations based on these measurements reveal the relative abundances, frequencies, and distributions of the plant species that compose the community.

Procedure 32.2
Assessing a community with the line-intercept method

1. With the help of your instructor, locate a suitable field site with a plant community to be examined.
2. Obtain a measuring tape 10 to 30 meters long, a meter stick, and a notepad. If a measuring tape is unavailable, use a measured piece of string or rope.
3. Assess the general layout of the community to be sampled. With the aid of your instructor decide on a reasonable set of criteria to govern the placement of a transect for each group of students. You may establish a baseline along one side or within the community. Points along this line can be used to randomly establish starting points for the transects used to measure plant distributions.

Question 1
a. What concepts or ideas should govern the placement of your transect to obtain a representative sample of the community? _____
_____

b. Are there any "wrong" places to put a transect? _____
Why or why not? _____
_____

TABLE 32.1

SUMMARY OF RAW DATA FOR SPECIES OCCURRING ALONG A TRANSECT

Transect Length _____   Number of Intervals _____

Species$_i$	Total Number of Individuals Encountered	Total Length of Transect Intercepted	Number of Intervals in Which Species$_i$ Occurs
	$\Sigma=$	$\Sigma=$	$\Sigma=$

TABLE 32.2

OCCURRENCE OF EACH SPECIES IN A SELECTED COMMUNITY USING PARAMETERS OF THE LINE-INTERCEPT METHOD

Species$_i$	Density	Relative Density	Coverage	Relative Coverage	Frequency	Relative Frequency	Importance Value

4. You and your lab partners will work on a single transect. Stretch the measuring tape on the ground to establish a transect.

5. Divide the transect into 1 or 5 m intervals. Each interval represents a separate unit of the transect.

6. At one end of the transect begin counting plants that touch, overlie, or underlie the transect line. For each plant encountered record the type of plant (species) and the length of the line that the plant intercepts. For plants that overhang the line record the length of the imaginary vertical plane of the line that the plant would intercept. Record this raw data for each interval in your field notepad. Also record any uncovered (bare) lengths within the transects.

7. When all plants from all intervals have been recorded, summarize your data in table 32.1.

8. Use the data in table 32.1 to calculate the following parameters for each species within the community. Record your results in table 32.2.

$Density_i = n_i/length_t$

where n_i is the total number of individuals of species i encountered in all intervals, and $length_t$ is the total length of the transect.

$Relative\ density_i = n_i/Sn$

where Sn is the total number of all individuals encountered for all species.

$Coverage_i = length_i/length_t$

where $length_i$ is the sum of the intercept lengths for all occurrences of species i.

Question 2

How would you calculate "relative coverage" as opposed to "coverage"? Record your answer below.

$Relative\ coverage_i =$

$Frequency_i = intervals_i/intervals_t$

where $intervals_i$ is the number of intervals containing species i and $intervals_t$ is the total number of intervals.

Relative frequency = $frequency_i/frequency_t$

where $frequency_i$ is the frequency for species i (calculated previously) and $frequency_t$ is the total of frequencies for all species.

Importance value = Relative density + Relative coverage + Relative frequency

Question 3

a. What is the meaning of an importance value? _____

_____

b. Why would we calculate this in addition to density, coverage, and frequency? _____

_____

_____

PLANT INTERACTIONS

Competition

Competition is the interaction among individuals seeking a common resource that is scarce. The interaction is always negative (disadvantageous) for at least one of the competitors. The intensity of competition for a resource such as light, food, water, space, or nutrients depends on the amount of the resource, the number of individuals competing, and the needs of each individual for that resource. Competition by members of the same species is **intraspecific competition,** whereas competition by members of different species is **interspecific competition.**

Procedure 32.3

Examine competition by sunflower seedlings

1. Obtain five pots containing enough potting soil to plant seeds. Measure the area at the surface of the potting soil.

2. Follow your lab instructor's directions to plant 2, 4, 8, 16, and 32 sunflower seeds in the five pots. In each of the pots plant at least 20% more seeds than the number specified, to be reduced later to the appropriate number. Label each pot with the number of seeds, the date, and your name.

3. Water each of the pots gently with a consistent amount of water.

4. Place the pots in the greenhouse so that each group of seedlings is exposed to the same environmental conditions (e.g., light, temperature).

5. Examine the pots at regular intervals as directed by your instructor. Record measurements for the parameters listed in table 32.3.

Question 4

a. Was competition greater in the more crowded pots?

_____

b. Which parameters best demonstrated the effects of

competition? _____

c. What other characteristics of the competitors might

you have measured? _____

d. Did competition more noticeably affect the number of individuals or the biomass of each individual?

_____

e. In what environments would you expect that competition among sunflowers would be most intense?

_____

f. Would you expect different results if different potting

soil was used? _____

Why? _____

_____

g. In any given environment, would you expect inter- or intraspecific competition to be the most intense?

_____

Why? _____

_____

h. Would plants and animals compete for the same

resources? _____

What might be some differences in how plants and

animals compete? _____

_____

TABLE 32.3

EFFECTS OF COMPETITION ON SUNFLOWER SEEDLINGS

	Plants per Pot				
	2	4	8	16	32
General Observations					
Date _____					
Date _____					
Date _____					
Date _____					
Mean Height of Individuals					
Date _____					
Date _____					
Date _____					
Date _____					
Range of Height of Individuals					
Date _____					
Date _____					
Date _____					
Date _____					
Mean Width of Ten Widest Leaves					
Date _____					
Date _____					
Date _____					
Date _____					
Mean Fresh Weight of Aboveground Biomass					
Date _____					

Allelopathy

Allelopathy is the inhibition of a plant's germination or growth by exposure to chemicals produced by another plant. These chemicals may be airborne or leach from various plant parts. Rainfall, runoff, and diffusion typically distribute inhibitory chemicals in the immediate area surrounding the producing plant.

Question 5

a. What obvious advantage does a plant that produces such chemicals have? _____

_____

b. What possible disadvantages might there be? _____

_____

TABLE 32.4

EFFECTS OF ALLELOPATHY ON GERMINATION AND GROWTH RATE OF SEEDS

Plant Extract _____   Seed Species _____   Total No. of Seeds _____

Time	% Germinated	Length of Radicle	Observations
Rep 1			
Rep 2			
Rep 3			

Plant Extract _____   Seed Species _____   Total No. of Seeds _____

Time	% Germinated	Length of Radicle	Observations
Rep 1			
Rep 2			
Rep 3			

Plant Extract _____   Seed Species _____   Total No. of Seeds _____

Time	% Germinated	Length of Radicle	Observations
Rep 1			
Rep 2			
Rep 3			

Plant Extract _____   Seed Species _____   Total No. of Seeds _____

Time	% Germinated	Length of Radicle	Observations
Rep 1			
Rep 2			
Rep 3			

Procedure 32.4
Demonstrate allelopathy

1. Determine from your instructor the overall experimental design for the class; that is, how many plants your group will test and how many replicates you will set up for each plant.

2. Obtain tissue (stems and leaves) from the variety of plants provided by your instructor. Some of these plants are suspected to be inhibitory.

3. For each plant: Mix 10 g of tissue with 100 mL of water in a blender and homogenize. Let the slurry soak for 5 to 10 minutes to leach chemicals from the disrupted tissue.

4. Filter or strain the slurry to remove large particulates. Drain the filtrate into a beaker.

5. Obtain a petri plate and line its bottom with a circular piece of filter paper.

6. Label the petri plate appropriately for the plant extract and replicate being tested in that plate.

7. Saturate the filter paper with a measured amount (5–8 mL) of the extract.

8. Repeat steps 3 to 7 for each replicate and each plant being tested.

9. Obtain seeds of radish, lettuce, or oat. Distribute fifty seeds uniformly on the filter paper in each dish.

10. Your instructor may enhance the experimental design by asking you to set up replicate plates for each extract and to test the effects on different kinds of seeds. Follow his or her instructions.

11. Incubate the covered dishes at room temperature in the laboratory or in the greenhouse. After 24 and 48 hours count the number of seeds that have germinated and calculate the percent germinated.

12. After 72 hours (or the length of time specified by your instructor) measure the length of the radicle and make relevant observations about the apparent rate of growth of the emerged embryos. Extend your observations for as many days as specified by your instructor.

13. Record your results in table 32.4.

Question 6
a. Did your observations of growth rate reveal any differences in allelopathy among tested plants?

_____

b. Which plant species demonstrated the most intense allelopathy? _____

32-6

Summary of Activities

1. Characterize a terrestrial community with a variety of observations.

2. Complete a line-intercept assessment of plant abundance.

3. Experimentally determine the effects of competition on sunflower seedlings.

4. Assess the allelopathic effects of a variety of plant extracts on seed germination and growth.

Questions for Further Thought and Study

1. Interactions in some plant communities are more complicated than others. What factors add to a community's complexity?

2. What characteristics of a community make it more resilient than other communities after a disturbance?

3. What characteristics would indicate that a community has not been disturbed for a long time?

4. How does competition influence natural selection?

Doing Botany Yourself

Design and conduct an experiment that determines which organs of an allelopathic plant—that is, leaves, stems, or roots—produce the most active inhibitors. Is there a selective advantage to some tissue producing more inhibitors than another tissue?

a p p e n d i x   a

Measurements in Science

THE METRIC SYSTEM

The metric system is a standardized system of measurement used by scientists throughout the world. It is also the measurement system used in everyday life in most countries. Although the metric system is the only measurement system ever acknowledged by Congress, the United States remains out of step with the rest of the world in clinging to the antiquated English system of measurements involving pounds, inches, ounces, and so on.

Metric units commonly used in biology include the following:

> meter (m)—the basic unit of length
>
> liter (L)—the basic unit of volume
>
> gram (g)—the basic unit of mass
>
> degree Celsius (°C)—the basic unit of temperature

Unlike the English system with which you are already familiar, the metric system is based on units of ten, thus simplifying interconversions. This base-ten system is similar to our monetary system, in which 10 cents equals a dime, 10 dimes equals a dollar, and so on. Units of ten in the metric system are indicated by Latin and Greek prefixes placed before the words for the base units (see table A.1). Thus, multiply by

0.01	to convert	cg to g
0.1	to convert	dm to m
1,000	to convert	kg to g
0.000001	to convert	mμ to g
0.1	to convert	mm to cm

For example,
620 g = 0.620 kg = 620,000 mg = 6,200 dg = 62,000 cg.

UNITS OF LENGTH

The **meter** is the basic unit of length.

> **1 m = 39.4 inches (in) = 1.1 yards (yd)**
> **1 in = 2.54 cm**
> **1 km = 1,000 m = 10^3 m = 0.62 miles (mi)**
> **1 ft = 30.5 cm**
> **1 cm = 0.01 m = 10^{-2} m = 0.39 in = 10 mm**
> **1 yd = 0.91 m**
> **1 nm = 10^{-9} m = 10^{-6} mm = 10 angstroms (Å)**
> **1 mi = 1.61 km**
> **470 m = 0.470 km**

Units of area are squared (two-dimensional) units of length.

> **1 m^2 = 1.20 yd^2 = 1,550 in^2 = 1.550×10^3 in^2**
> **= 10.8 ft^2**
>
> **1 km^2 = 0.39 mi^2**
>
> **1 cm^2 = 100 mm^2 = 0.16 in^2**
>
> **1 hectare = 10,000 square meters (m^2) = 2.47 acres**

The following comparisons will help you appreciate these conversions of length and area:

Height of the Washington Monument	169.1 m	555 ft
Length of a housefly	0.5 cm	0.2 in
Distance kangaroos can hop in one leap	7.6 m	25 ft
Diameter of a pore in a hen's egg	17 μm	0.0007 in
Surface area of a flu virus	10^{-8} mm^2	1.6×10^{-11} in^2
Skin area of average-sized woman	1.58 m^2	17 ft^2
Surface area of U.S. paper currency	104 cm^2	16.1 in^2

Measurements of area and volume can use the same units.

METRIC SYSTEM DIVISIONS AND MULTIPLES OF THE BASIC UNIT OF LENGTH, THE METER

Prefix (Latin)	Division of Metric Unit (meter)		Equivalent	Prefix (Greek)	Multiple of Metric unit (Meter)		Equivalent
deci (d)	0.1	10^{-1}	tenth part	deka (da)	10	10^1	tenfold
centi (c)	0.01	10^{-2}	hundredth part	hecto (h)	100	10^2	hundredfold
milli (m)	0.001	10^{-3}	thousandth part	kilo (k)	1,000	10^3	thousandfold
micro (μ)	0.000001	10^{-6}	millionth part	mega (M)	1,000,000	10^6	millionfold
nano (n)	0.000000001	10^{-9}	billionth part	giga (G)	1,000,000,000	10^9	billionfold
pico (p)	0.000000000001	10^{-12}	trillionth part	teta (T)	1,000,000,000,000	10^{12}	trillionfold

$$1 \text{ m}^3 = 35.3 \text{ ft}^3 = 1.31 \text{ yd}^3$$

$$1 \text{ cm}^3 \text{ (cc)} = 0.000001 \text{ m}^3 = 0.061 \text{ in}^3$$

The following comparisons will help you appreciate these conversions of volume:

Soda can	355 mL	12.0 oz
Legal-sized filing cabinet	0.4 m^3	14 ft^3
Sugar cube	1 cm^3	0.061 in^3

UNITS OF MASS

The **gram** (g) is the basic unit of mass.

$$1 \text{ g} = \text{mass of 1 cm}^3 \text{ of water at } 4°C = 0.035 \text{ oz}$$

$$1 \text{ kg} = 1,000 \text{ g} = 10^3 \text{ g} = 2.2 \text{ lb}$$

$$1 \text{ mg} = 0.001 \text{ g} = 10^{-3} \text{ g}$$

The following comparisons will help you appreciate these conversions of mass:

Penicillin molecule	10^{-18} kg	2×10^{-18} lb
Giant amoeba	10^{-8} kg	2×10^{-8} lb
Human	10^2 kg	220 lb
Nickel	5 g	0.176 oz

Remember that mass is not necessarily synonymous with weight. Mass measures an object's potential to interact with gravity, whereas weight is the force exerted by gravity on an object. Thus, a weightless object in outer space has the same mass as it has on earth.

UNITS OF VOLUME

The **liter** (L) is the basic unit of volume. A thermos bottle holds about 1 liter; a standard flush toilet flushes about 20 liters of water. Units of volume are cubed (three-dimensional) units of length.

$$1 \text{ liter} = 1,000 \text{ cm}^3 = 1,000 \text{ mL} = 0.001 \text{ m}^3$$

$$= 2.1 \text{ pints} = 1.06 \text{ qt} \quad 1 \text{ cup} = 240 \text{ mL}$$

$$= 0.26 \text{ gal} = 1 \text{ dm}^3 = 0.001 \text{ m}^3$$

$$1 \text{ mL} = 0.035 \text{ fl oz}$$

UNITS OF TEMPERATURE

You are probably most familiar with temperature measured on the Fahrenheit scale, which is based on water freezing at 32°F and boiling at 212°F. Celsius temperatures are synonymous with centigrade temperatures, and these scales measure temperature in the metric system. Celsius (°C) temperatures are easier to work with than Fahrenheit temperatures since the Celsius scale is based on water freezing at 0°C and boiling at 100°C. You can interconvert temperatures in the two systems with the following formula:

5 × degrees Fahrenheit = (9 × degrees Celsius) + 160

For example:

45°C = 113°F (temperature at which any area of skin of an average person will feel pain)

40°C = 104°F (a hot summer day)

75°C = 167°F (hot coffee)

–5°C = 23°F (coldest area of freezer)

37°C = 98.6°F (human body temperature)

16°C = 60.8°F (coldest body temperature a person has been known to have and survive)

The Celsius temperature scale was developed in 1742 by Anders Celsius, a Swedish astronomer. Interestingly, Celsius originally set 0°C as the boiling point of water and 100°C as the freezing point of water. Soon thereafter,

HINTS FOR USING THE METRIC SYSTEM

1. Express measurements in units requiring only a few decimal places. For example, 0.3 m is more easily manipulated and understood than 300,000,000 nm.

2. When measuring water, the metric system offers an easy and common conversion from volume measured in liters to volume measured in cubic meters to mass measured in grams: 1 mL = 1 cm³ = 1 g.

3. Familiarize yourself with manipulations within the metric system. Work within one system; do not convert back and forth between the metric and English systems.

4. The metric system uses symbols rather than abbreviations. Therefore, do not place a period after metric symbols (e.g., 1 g, not 1 g.). Use a period after a symbol only at the end of a sentence.

5. Do not mix units or symbols (e.g., 9.2 m, not 9 m 200 mm).

6. Metric symbols are always singular (e.g., 10 km, not 10 kms).

7. Except for degree Celsius, always leave a space between a number and a metric symbol (e.g., 20 mm, not 20mm).

8. Use a zero before a decimal point when the number is less than one (e.g., 0.42 m, not .42 m).

J. P. Christine revised the scale to its present-day form, with 0°C as the freezing point and 100°C as the boiling point of water. The Celsius scale is identical to the centigrade scale.

The following comparisons will help you appreciate these conversions of temperature:

Water freezes	0°C	32°F
Butter melts	31°C	87°F
Water boils	100°C	212°F
Fireplace fire	827°C	1,520°F

A-3

appendix b

Spectrophotometry: A Lab Exercise

Objectives

By the end of this exercise you will be able to

1. operate a spectrophotometer;
2. describe the parts of a spectrophotometer and the function of each;
3. construct absorption spectra for cobalt chloride and chlorophyll;
4. construct and use a standard curve to determine the unknown concentration of a dissolved chemical.

Absorption and reflection of different wavelengths of lights are part of your everyday experience. Different objects have different colors. When you recognize things with color, you are observing that they absorb and reflect different wavelengths of light. Light includes the visible wavelengths of the electromagnetic spectrum, which is only a small part of the total spectrum. The entire electromagnetic spectrum includes radiation with wavelengths from less than 1 to more than 1 million nanometers. Visible light only represents wavelengths between 380 and 750 nm (fig. B.1) (see plate 1, fig. 1).

Chlorophyll is a colorful and well-known pigment. What color of light do you think is least effective for chlorophyll's role of absorbing light for photosynthesis? The answer is green because chlorophyll reflects green light and absorbs the energy of other colors. Similarly, other molecules absorb certain colors (visible wavelengths) and reflect other colors.

Question 1
a. What biologically important molecules other than chlorophyll absorb and reflect certain colors?
b. Is the absorbance of light critical to their function or just a consequence of their molecular structure?

SPECTROPHOTOMETER

Biologists routinely determine the presence and concentration of dissolved chemicals such as phosphates that may pollute lake water. **Spectrophotometry** is one of our most versatile and precise techniques for chemical assays. Spectrophotometry is based on the principle that every different atom, molecule, or chemical bond absorbs a unique pattern of wavelengths of light. For example, the nitrogenous base cytosine absorbs a different pattern of light than does adenine, uracil, or any other molecule with a different structure. As part of this pattern, some wavelengths are absorbed and some are not. Conversely, a unique pattern of light is reflected by each chemical. We use an instrument called a **spectrophotometer** to measure the amount of light absorbed and transmitted by a dissolved chemical. With dissolved chemicals we usually refer to the nonabsorbed light as transmitted rather than reflected. With this measurement of absorbance or transmittance we can identify a chemical and its concentration.

A spectrophotometer is an instrument that shines a known and specific wavelength of light into a solution and measures how much light the solution absorbs and how much it transmits. The basic parts of a spectrophotometer are shown in figure B.2.

The **light source** of a spectrophotometer produces white light, which is a combination of all visible wavelengths. A mixture of all colors of light is white. Spectrophotometry may involve wavelengths outside the visible range, such as ultraviolet and infrared, but the spectrophotometer must have special light sources to produce these wavelengths. In this lab exercise you will work only with visible light.

The **filter** is adjusted to select the wavelength that you wish to pass through the sample. The filter may be a prism that separates white light into a rainbow of colors

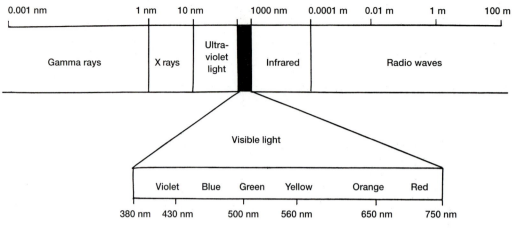

FIGURE B.1

The electromagnetic spectrum. Light is a form of electromagnetic energy and is conveniently thought of as a wave. The shorter the wavelength of light, the greater the energy. Visible light represents only the small part of the electromagnetic spectrum between 380 and 750 nanometers (nm). (UV stands for ultraviolet light.)

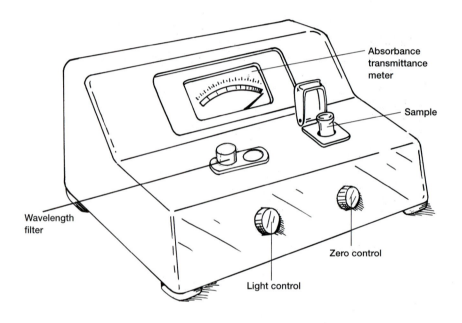

FIGURE B.2

Spectrophotometer.

and focuses the desired wavelength (color) on the sample. Or the filter may be a series of colored glass plates that absorb all but the selected wavelength. The selected wavelength is focused on the sample.

The **sample** is a solution contained in a clear test tube or cuvette made of glass or quartz. Light of the specific wavelength determined by the filter passes into the sample, where it may be completely or partially absorbed or transmitted. The amount of light absorbed depends on the amount and kind of chemicals in the solution.

A **blank** is a test tube containing only the solvent used to dissolve the chemical you are analyzing. A blank is used to calibrate the spectrophotometer for the solution used in your experiment. For most of your experiments, the solvent, and therefore the blank, is distilled water.

Any light transmitted through the sample exits the sample on the opposite side and is focused on a photodetector, which converts light energy into electrical energy. The amount of electricity produced by the photodetector is proportional to the amount of transmitted light; the more light transmitted, the more electricity produced. The meter on the front of the machine measures the electrical current produced by the photodetector and displays the results on a scale of absorbance or transmittance values.

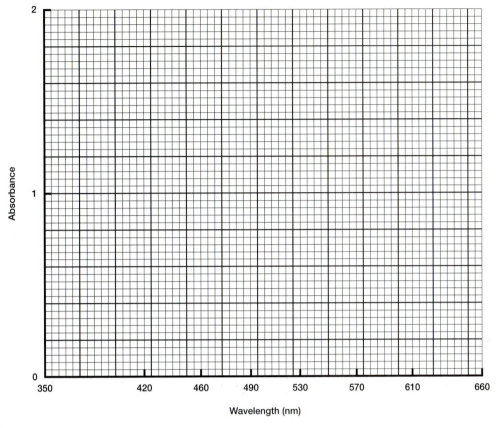

Absorption spectrum of cobalt chloride

FIGURE B.3

Absorption spectrum of cobalt chloride.

ABSORPTION SPECTRUM OF COBALT CHLORIDE

Your first task is to learn to operate the spectrophotometer while deriving the absorption spectrum of a common chemical, cobalt chloride ($CoCl_2$). The pattern of wavelengths absorbed by $CoCl_2$ is its "fingerprint" since it is unique to that chemical. This fingerprint, or absorption spectrum of the chemical, is represented as a graph relating absorbance to wavelength (fig. B.3).

Procedure B.1
Determine the absorption spectrum of $CoCl_2$

1. Turn on the spectrophotometer. Let it warm up for 10 to 15 minutes before you begin work.

2. Prepare seven solutions in spectrophotometer tubes (test tubes or cuvettes) with the mixtures of distilled water and stock solution of $CoCl_2$ (100 mg/mL) listed in table B.1. Only tubes 0 and 6 are needed to determine the absorption spectrum. You will use the others later in the lab period.

3. After you have prepared the dilutions clean the outside of all the tubes with a cloth or paper towel.

Question 2
Why is cleaning the sample tubes important?

4. Cap the tubes and label the caps 0 to 6. If you label the tube rather than the caps, then be sure that the labels don't interfere with light entering the tube while it is in the spectrophotometer.

5. Place the blank (tube 0) in the sample holder of the spectrophotometer.

6. Adjust the filter to the lowest wavelength (350 nm) and read the absorbance value indicated on the meter. The distilled water blank has no color and should not absorb any visible light.

7. If the absorbance is not zero, use the zero-control knob to calibrate the meter to zero on the absorbance scale.

8. Remove the blank and replace it with tube 6 (50 mg/mL). This is the sample you will use to determine the absorption spectrum of cobalt chloride.

TABLE B.1

VOLUMES OF COBALT CHLORIDE STOCK SOLUTION (100 MG/ML) AND WATER FOR PREPARING SEVEN KNOWN DILUTIONS

Tube Number	Concentration of $CoCl_2$ (mg/mL)	$CoCl_2$ Stock (mL)	Distilled H_2O (mL)
0	0	0	10.0
1	1	0.1	9.9
2	10	1.0	9.0
3	20	2.0	8.0
4	30	3.0	7.0
5	40	4.0	6.0
6	50	5.0	5.0

TABLE B.2

ABSORBANCE VALUES FOR $CoCl_2$ (50 MG/ML)

Wavelength	Absorbance	Wavelength	Absorbance
350 nm	_____	530 nm	_____
420 nm	_____	570 nm	_____
460 nm	_____	610 nm	_____
490 nm	_____	660 nm	_____

9. After the meter has stabilized (5–10 seconds) read the absorbance value and record it in table B.2.

10. Remove tube 6 and adjust the filter to 420 nm.

11. Put the blank back into the spectrophotometer and readjust for zero absorbance at the new wavelength. The spectrophotometer should be recalibrated with the blank often, especially when you change the wavelength.

12. Insert tube 6 and measure its absorbance at the new wavelength. Record the absorbance in table B.2.

13. Complete table B.2 for the other wavelengths by repeating steps 5 to 12 and measuring the absorbance of the contents of tube 6.

The absorbance values in table B.2 represent the absorption spectrum for $CoCl_2$ and are expressed best with a graph. Plot on figure B.3 the absorbance values recorded in table B.2 for each wavelength. Connect the points with straight lines.

Question 3

a. What wavelength represents the peak absorbance of $CoCl_2$?

b. Would you expect a curve of the same shape for another molecule such as chlorophyll? Why or why not?

ABSORBANCE AND THE STANDARD CURVE

The double scale on the spectrophotometer indicates that you can measure either absorbance or transmittance of radiation. **Absorbance** is the amount of radiation retained by the sample, and **transmittance** is the amount of radiation passing through the sample. In mathematical terms, transmittance is the amount or intensity of light exiting the sample divided by the amount entering the sample. Transmittance is usually expressed as a percent:

Percent transmittance = $(I_t/I_o) \times 100$

where I_t = transmitted (exiting) light intensity, and I_o = original (entering) light intensity.

Absorbance is not simply the reciprocal of transmittance. Rather, absorbance is the logarithm of the reciprocal of transmittance and is expressed as a ratio with no units:

Absorbance = $\log_{10} (I_o/I_t)$

You may ask why scientists use a logarithm rather than some other mathematical transformation to calculate absorbance. The answer is that when absorbance is calculated as a logarithm, it is directly proportional to the

B-4

TABLE 5.3

ABSORBANCE VALUES FOR SIX SOLUTIONS OF KNOWN CONCENTRATION (STANDARDS) OF CoCl₂ AT THE PEAK ABSORBANCE WAVELENGTH

Concentration of Standards (mg CoCl₂/mL)	Peak Wavelength = _____ Absorbance
1	_____
10	_____
20	_____
30	_____
40	_____
50	_____

concentration of the substance in solution. Therefore, a twofold increase in absorbance indicates a twofold increase in concentration. This convenient and direct relationship between concentration and absorbance, known as the Beer-Lambert law, helps scientists measure an unknown concentration of a chemical.

A graph showing a chemical's concentration versus its absorbance of a wavelength of light is called a **standard curve,** and the relationship is expressed as a straight line. In this exercise, you will construct a standard curve and then use it to determine some unknown concentrations of solutions of CoCl₂ prepared by your instructor. Use the six dilutions that you prepared previously to construct your standard curve for CoCl₂. These solutions are **standards** because their concentrations are known and they are used to determine the concentration of an unknown solution. The absorbance of each standard is measured at the peak wavelength of the absorption spectrum for CoCl₂.

Procedure B.2
Construct a standard curve for cobalt chloride

1. Refer to your data in table B.2. Determine the wavelength of peak absorbance for CoCl₂, and set the filter of your spectrophotometer to this wavelength.

2. Insert the solvent blank (tube 0) and adjust the spectrophotometer for zero absorbance.

3. Replace the blank with tube 1 (1 mg CoCl₂/mL), measure its absorbance, and then record the absorbance value in table B.3.

4. Repeat steps 2 to 3 for the other five tubes (standards). Be sure and check the zero-absorbance calibration with the blank before each standard measurement.

To construct your standard curve for CoCl₂, plot the data in table B.3 with *Concentration (mg/mL)* on the horizontal axis

and *Absorbance* on the vertical axis. (Use the graph paper at the end of appendix B.) Because the relationship between concentration and absorbance is linear (directly proportional), draw one straight line that lies as close as possible to each data point. Do not merely connect the dots.

Question 4
Do the plotted data points of your standard curve lie on a straight line? _____

Using the Standard Curve to Measure the Unknown Concentration of a Solution

After you have created a standard curve, measuring the unknown concentration of a CoCl₂ solution is easy. Your instructor has prepared a series of numbered tubes containing unknown concentrations of CoCl₂.

Procedure B.3
Determine unknown concentrations

1. Obtain a tube with an unknown solution and record the tube number in table B.4.

2. Use the blank (tube 0) to zero the spectrophotometer at the wavelength of peak absorbance for CoCl₂.

3. Measure the absorbance of the unknown solution and record this value in table B.4.

4. Find this absorbance value on the vertical axis of your standard curve and draw a line from this point parallel to the horizontal axis until the line intersects the standard curve (fig. B.4).

5. Draw a line from the intersection straight down until it intersects the horizontal axis. This point on the horizontal axis marks the concentration of the unknown solution.

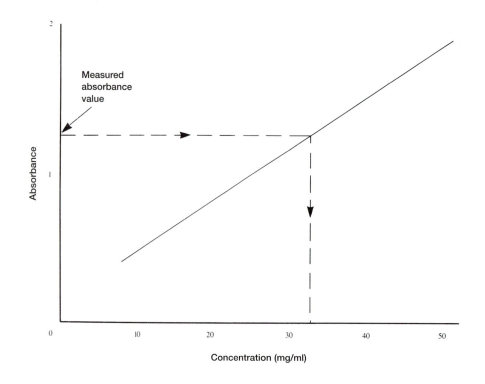

FIGURE B.4

A standard curve showing the graphical determination of the concentration of an unknown solution of $CoCl_2$.

MEASUREMENTS OF ABSORBANCE AND CONCENTRATION FOR FOUR UNKNOWN SOLUTIONS OF $CoCl_2$

Tube Number	Absorbance	Concentration (mg/mL)
Unknown _____	_____	_____
Unknown _____	_____	_____
Unknown _____	_____	_____
Unknown _____	_____	_____

6. Record the concentration of the unknown solution in table B.4.

7. Obtain three more tubes with unknown solutions, determine their absorbance and concentration, and record the values in table B.4. Ask your instructor to check your results.

ABSORPTION SPECTRUM OF CHLOROPHYLL

To give you more experience with absorption spectra, your instructor has prepared a plant extract containing chlorophyll, the photosynthetic pigment (see plate I, fig. 2). The extract was made by grinding leaves in acetone.

CAUTION

Acetone is flammable. Keep all solvents away from hotplates and flames at all times.

Procedure B.4
Determine the absorption spectrum of chlorophyll

1. Obtain a tube of the extract and prepare a blank. Your instructor will provide the solvent used for the blank.

2. Using the previous procedure for determining an absorption spectrum, measure the absorbance of chlorophyll for at least eight wavelengths available on your spectrophotometer, as listed in table B.5.

3. Record your results in table B.5.

4. Graph your results (in fig. B.5) as you did for the absorption spectrum of $CoCl_2$.

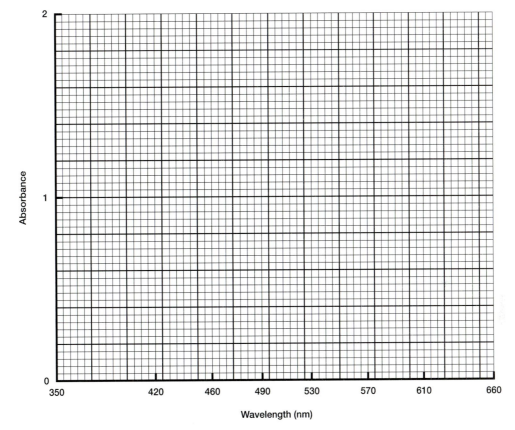

Absorption spectrum of chlorophyll

FIGURE B.5

Absorption spectrum of chlorophyll.

Question 5

a. What is the proper blank for determining the absorption of chlorophyll in a plant extract? _____

_____

b. Which wavelengths are least absorbed by chlorophyll?

_____

c. Which wavelengths are most absorbed by chlorophyll?

_____

Questions for Further Thought and Study

1. How do you effectively separate the absorption spectrum of the solvent from that of the solute to ensure accuracy?

2. Ask your instructor how the chlorophyll solution that you used to determine an absorption spectrum was prepared. What is the difference between a plant extract and chlorophyll?

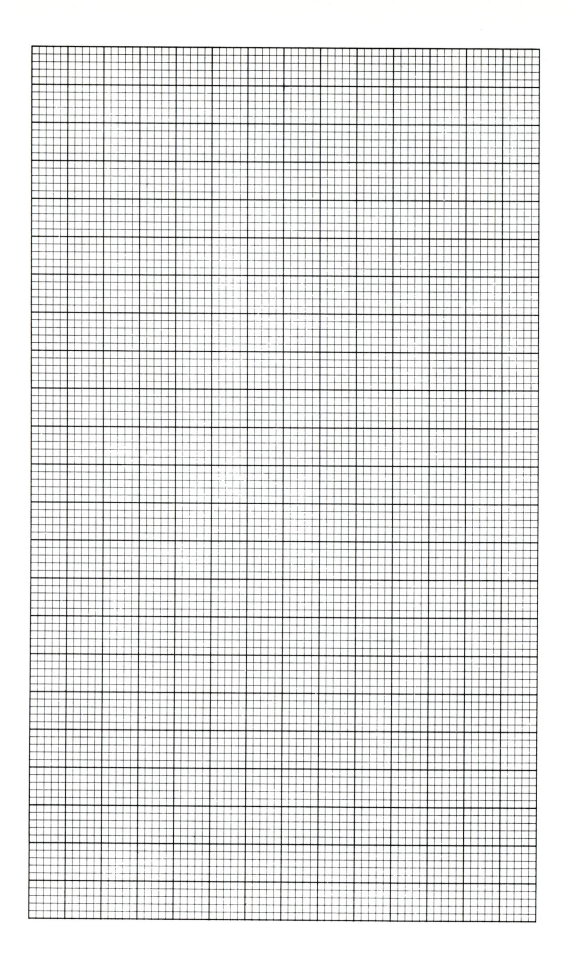

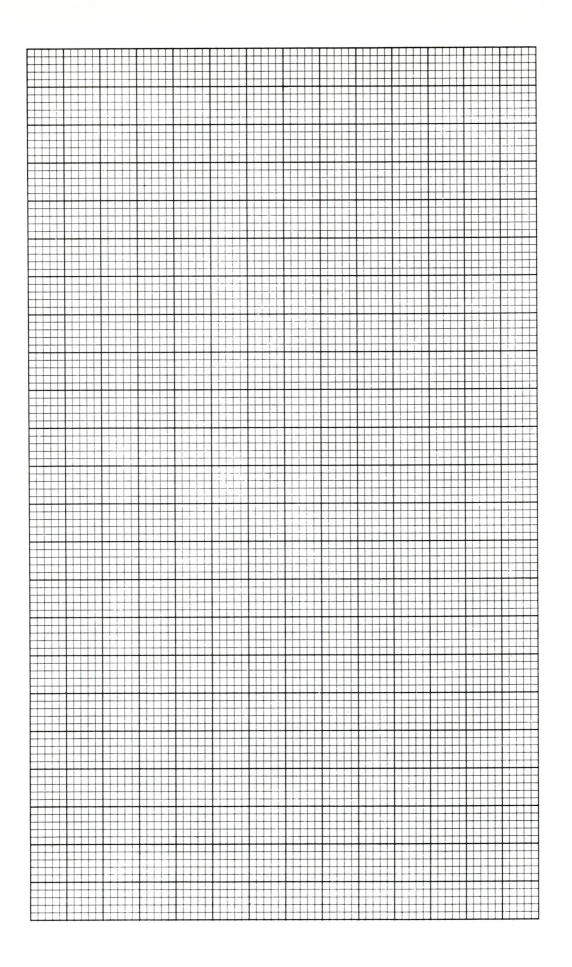

appendix c

How to Write a Scientific Paper

"I have learned that when I write a research paper I do far more than summarize conclusions already neatly stored in my mind. Rather, the writing process is where I carry out the final comprehension, analysis, and synthesis of my results."

—Sidney Perkowitz

Your instructor may occasionally ask you to submit written reports describing the work you did in lab. Although these reports will not be published in scientific magazines or journals, they are nevertheless important because they will help you learn to write a scientific paper. A **scientific paper** is a written description of how the scientific method was used to study a problem.

Understanding how to write a scientific paper (such as a lab report) is important for several reasons. In general, scientists become known (or remain unknown) by their publications in books, magazines, and scientific journals. Regardless of the potential importance of a scientist's discoveries, poor writing delays or prohibits publication since it makes it difficult to understand what the scientist did or the importance of the work. Poor writing usually indicates an inability or unwillingness of a scientist to think clearly. More important, scientific papers are the vehicle for the transmission of scientific knowledge; they are available for others to read, test, refute, and build on. Few skills are more important to a scientist than learning how to write a scientific paper.

THE PARTS OF A SCIENTIFIC PAPER

Before reading this appendix go to the library and browse through a few biological journals such as *American Journal of Botany*, *Ecology*, or *Journal of Cell Biology*. Make photocopies of one or two of the articles that interest you. As you'll see, scientific papers follow a standard format that reflects the scientific method. Almost all scientific papers have these parts:

Title

List of authors

Abstract

Introduction

Materials and methods

Results

Discussion

References

Understanding this format eases the burden of writing a scientific paper, since writing is really an exercise in organization. Refer to the journal articles you photocopied in the library as you read this appendix.

Title

The title of a paper is a short label (usually less than ten words) that helps readers quickly determine their interest in the paper. The title should reflect the paper's content and contain the fewest number of words that adequately express the paper's content. The title should never contain abbreviations or jargon. Jargon is overly specialized or technical language.

List of Authors

Only those who actively contributed to the design, execution, or analysis of the experiment should be listed as authors.

Abstract

The abstract is a short paragraph (usually less than 250 words) that summarizes (1) the objectives and scope of the problem, (2) methodology, (3) data, and (4) conclusions. The abstract contains no references.

Introduction

The introduction concisely states why you did the work. Avoid exhaustive reviews of what has already been published; rather, limit the introduction to just enough pertinent information to orient the reader to your study. The introduction of a scientific paper has two primary parts. The first part is a description of the nature and background of the problem. For example, what do we already know (or not know) about the problem? This description is developed by citing other scientists' work, to give a history of

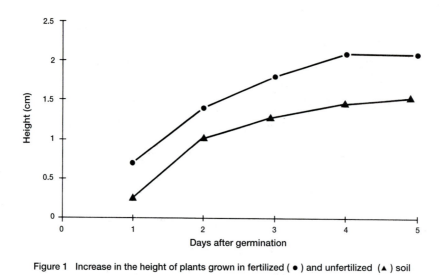

Figure 1 Increase in the height of plants grown in fertilized (●) and unfertilized (▲) soil

FIGURE C.1

Sample graph from a scientific paper.

the study of the problem, and to point out gaps in our knowledge. The second part of the introduction states the objectives of the study.

Materials and Methods

The materials and methods section describes how, when, where, and what you did. It should contain enough detail to allow another scientist to repeat your experiment, but it should not be overwhelming. **Materials** include items such as growth conditions, organisms, and the chemicals used in the experiment. Avoid trade names of chemicals, and describe organisms with their scientific names (e.g., *Zea mays* rather than "corn"). Also describe growth conditions, diet, lighting, temperature, and so on. **Methods** are usually presented chronologically and subdivided with headings. Examples of methods include sampling techniques, types of microscopy, and statistical analyses. If possible, use references to describe methods.

Experiments described in a scientific paper must be reproducible. Thus, the quality of the materials and methods section of a paper is judged by the reader's ability to repeat the experiment. If a colleague can't repeat your experiment, the materials and methods section is probably poorly written.

For most lab reports, do not copy the experimental procedures word for word from the lab manual. Rather, summarize what you did in several sentences.

Results

The results section is the heart of a scientific paper. It should clearly summarize your findings and leave no doubt about the outcome of your study. For example, state that

"All plants died within 4 days of being treated with herbicide" or "Table 1 shows the influence of 2,4-D on leaf growth." Keep it simple and to the point.

Tables and graphs are excellent ways to present results but shouldn't completely replace a written summary of results. Tables are ideal for large amounts of numerical data, and graphs are an excellent way to summarize data and show relationships between independent and dependent variables. The variable that the scientist established and controlled during the experiment is the **independent variable.** It is presented on the x-axis of the graph. The height of a plant might be an independent variable in an experiment measuring how fertilizer affects growth (fig. C.1). Similarly, time and temperature are often independent variables. The **dependent variable** changes in response to changes in the independent variable and is presented on the y-axis of the graph. Weight and growth rate are examples of dependent variables that may change in response to light, temperature, diet, and so on. Graphs must also have a title (e.g., "Influence of Temperature on Root Elongation"), labeled axes (e.g., "Temperature," Root Elongation"), and scaled units along each axis appropriate to each variable (e.g., °C, mm h^{-1}). When submitting a scientific paper, place each table and graph on a separate page.

Discussion

It's not enough to simply report your findings; you must also discuss what they mean and why they're important. This is the purpose of the discussion section of a scientific paper. This section should interpret your results relative to the objectives you described in the introduction and

C-2

answer the question "So what?" or "What does it mean?" A good discussion section should do the following:

- Discuss your findings; that is, present relationships, principles, and generalizations.

- Point out exceptions and lack of correlations. Don't conceal anomalous results; rather, describe unsettled points.

- State how your results relate to existing knowledge.

- State the significance and implications of your data. What do your results mean? If your data are strong, don't hesitate to use statements beginning with "These results indicate that. . . ."

References

Scientists rely heavily on information presented in papers written by their colleagues. Indeed, the introduction, materials and methods, and discussion sections of a scientific paper often contain citations of other publications. The format for these citations varies in different biological journals. The following citation for an article is in the format recommended by the Council of Biology Editors:

White, H.B., III. Coenzymes as fossils of an earlier molecular state. J. Mol. Evol. 7:10–104; 1976.

A FEW SIMPLE RULES FOR WRITING EFFECTIVELY

Informative sentences and well-organized paragraphs are the foundation of a good scientific paper. Listed here are a few rules to help you write effectively. Following these rules won't necessarily make you a Hemingway, but it will probably improve your writing.

- Write clearly and simply. For example, "the biota exhibited a 100% mortality response" is a wordy and pretentious way of saying "all of the plants died." Remember, keep it simple and straightforward.

- Keep related words together. Consider the following sentence: "Atop the leaf, you perhaps can make out a small strand of vascular tissue." Do we really have to be atop the leaf to see the strand?! The author meant that "a small strand of vascular tissue is visible on the upper surface of the leaf."

- Prefer active voice. Write "Good writers avoid passive voice," not "The passive voice is avoided by good writers." Here are some other examples of passive voice:

  Poor: My first lab report will always be remembered. (passive)

Better: I'll always remember my first lab report. (active)

Poor: Examination of patients was accomplished by me. (Passive)

Better: I examined patients. (active)

- Write positively. For example, write "The plants were always sick" instead of "The plants were never healthy."

- Be definite and specific. For example, write "It rained every day for a week" instead of "A period of unfavorable growth conditions set in."

- Be sure of the meaning of every word you use, and write exactly what you mean. Refer to a dictionary and thesaurus to ensure clarity and proper word usage. For example:

  You allude to a book, and elude a pursuer.

  Something nauseous is sickening to contemplate, while a nauseated person is sick. (You can't be nauseous unless you're capable of making someone sick.)

- Delete unnecessary words. For example:

  advance notice = notice

  at this point in time = now

  be that as it may = but

  in the event that = if

  general consensus = consensus

  the question as to whether = whether

  young juvenile = juvenile

  student body = students

  due to the fact that = because

  chemotherapeutic agent = drug

- Use metric measurements (see appendix A).

- Be sure that each paragraph conveys one idea and has a topic sentence. The topic sentence should state the main idea of the paragraph.

- Have a friend or colleague read a draft of your writing and suggest improvements.

- Don't plagiarize. Learn to summarize, and be sure to cite all references from which you extracted information.

A neat and typed presentation is a must. If you use a word processor remember to spell check. Carefully proofread to catch mistakes. Put your work aside for at least 24 hours before you proofread.

If you're interested in learning more about improving your writing, read *The Elements of Style* (3d ed.) by W. Strunk and E.B. White (New York: Macmillan, 1979)

glossary

A

Absorption Spectrum A graph of the relative amount of light absorbed by a dissolved substance as a function of wavelength.

Accessory Pigment A light-absorbing pigment used in photosynthesis; typically transfers its energy to chlorophyll.

Adenosine Triphosphate (ATP) A nucleotide consisting of adenine, ribose, and three phosphate groups; serves as the energy currency of cellular metabolism.

Aerobe An organism requiring or tolerating molecular oxygen for its respiration.

Aerobic Requiring free oxygen; any biological process that can occur in the presence of free oxygen.

Allele One of two or more alternative states of a gene.

Allelopathy Inhibition of a plant by chemicals produced by another plant.

Alternation of Generations Reproductive cycle in which the gameophyte (haploid phase) produces gametes that fuse to form a diploid zygote. The zygote develops into the diploid sporophyte phase, which produces spores, which give rise to gametophytes.

Anaerobic Any process that can occur without oxygen, such as anaerobic fermentation; anaerobic organisms can live without free oxygen.

Anaphase The third stage of mitosis characterized by the separation of chromatids into two areas to become the new nuclei.

Antheridium Multicellular male reproductive structure of plants other than seed plants.

Apical Meristem The plant meristem tissue located at the tips of stems or roots.

Archegonium Sex organ that produces female gametes.

Autotrophic Able to synthesize all required nutrients from inorganic sources.

Auxin A plant hormone that controls cell elongation and other effects.

B

Bilateral Symmetry A body form in which the right and left halves of an organism are approximate mirror images of each other.

Bioassay Method to determine the unknown concentration of a physiologically active substance.

Brownian Movement Movement of small particles caused by collisions with constantly moving water molecules.

Budding A method of asexual reproduction consisting of mitosis with an unequal distribution of cytoplasm.

C

Cambium In vascular plants, embryonic tissue zones (meristems) that run parallel to the sides of roots and stems; consists of the vascular cambium and the cork cambium.

Catalyze To accelerate a chemical reaction by altering the activation energy of the reactants.

Cell Cycle Stages of a cell's activity, development, and replication.

Cell Plate Precursor of the cell wall, which separates the new cells produced in plant cell replication.

Cell Wall The rigid outermost layer of the cells of plants, fungi, bacteria, and some protists.

Centromere Condensed region eukaryotic chromosomes where sister chromatids are attached to one another after replication; composed of highly repeated DNA sequences.

Chitin A nitrogenous organic compound common in the exoskeleton of arthropods and the cell wall of some fungi.

Chlorophyll The green pigment of plant cells that is necessary for photosynthesis.

Chloroplast An organelle present in algae and plant cells that contains chlorophyll—the green pigment in plants that is the light receptor in photosynthesis.

Chromatid One of two daughter strands of a duplicated chromosome that are joined by a single centromere; separates and becomes a daughter chromosome at anaphase of the second meiotic division.

Chromoplast A plastid containing pigments other than chlorophyll.

Cilia A short, hairlike organelle on the surface of a cell; used for locomotion or movement of particles on the surface.

Collenchyma In plants, a supporting tissue composed of collenchyma cells; often found in regions of primary growth in stems and in some leaves. Collenchyma cells are elongated living cells with irregularly thickened primary cell walls.

Competition Interaction between organisms to obtain a required resource that is in limited supply.

Conidiophore An upright hypha bearing conidia.

Conidium An asexually produced fungal spore formed outside of a sporangium.

Conjugation The transfer of genetic material through a cytoplasmic bridge between cells; a common sexual process within some species of microorganisms.

Contractile Vacuole A cellular vacuole that collects and expels waste products.

Cotyledon Seed leaf; cotyledons generally store food in dicotyledons and absorb it in monocotyledons. The food is used during the course of seed germination.

Crossing-over The exchange of genetic material between homologous chromosomes during meiosis.

Cuticle A waxy or fatty noncellular layer on the outer wall of epidermal cells.

Cutin The substance composing the waxy or fatty noncellular layer on the outer wall of epidermal cells.

Cyclosis The flowing or streaming of cytoplasm within a cell.

Cytokinesis Separation of the cytoplasm into halves; usually associated with mitosis.

Cytoplasm The living matter within a cell excluding the nucleus.

D

Decomposer Organism that breaks down organic material into smaller molecules, which are then recirculated.

Denature To lose the native configuration of a protein or nucleic acid from excessive heat, extremes of pH, chemical modifications, or changes in solvent ionic strength or polarity; usually accompanied by loss of biological activity.

Depth of Field The thickness of an object that can be maintained in sharp focus.

Dialysis   Separation of substances in a solution by means of their unequal diffusion through a differentially permeable membrane.

Dichotomous Key   A written set of descriptive characteristics arranged in pairs and used to identify organisms.

Dicotyledon   A class of flowering plants generally characterized as having two cotyledons.

Differentially Permeable   Allowing selective passage of materials.

Differentiation   Developmental processes by which cells and tissues become specialized.

Diffusion   Net movement of particles from an area of high free energy to an area of low free energy.

Dioecious   Unisexual; having the male and female elements on different individuals of the same species.

Diploid   Having paired sets of chromosomes.

Double Fertilization   The fusion of the egg and sperm (resulting in a 2N fertilized egg, the zygote) and the simultaneous fusion of the second male gamete with the polar nuclei (resulting in a primary endosperm nucleus, which is often triploid, 3N).

E

Embryology   The stages and processes in the development of an embryo.

Energy of Activation   The inherent energy of reactants necessary for completion of a chemical reaction.

Enzyme   Protein molecule that catalyzes chemical reactions in living systems.

Enzyme-Substrate Complex   A combination of an enzyme and its substrate that stresses the chemical bonds of the substrate and promotes the enzymatic reaction.

Epidermis   The outermost layers of cells.

Eukaryotic   Having a distinct membrane-bound nucleus.

Eutrophication   Process whereby a body of freshwater becomes enriched with nutrients, increases in productivity, and accumulates organic debris.

Evolution   Genetic change in a population of organisms.

F

Fermentation   The enzyme-catalyzed extraction of energy from organic compounds without using oxygen; the enzymatic conversion, without oxygen, of carbohydrates to alcohols, acid, and carbon dioxide; the conversion of pyruvate to ethanol or lactic acid.

Fertilization   The fusion of the haploid nuclei of male and female gametes to form a diploid zygote.

Fitness   The genetic contribution of an individual to succeeding generations relative to the contributions of others.

Flagellum   A whiplike organelle composed of microtubules; used for locomotion by some microorganisms.

Fluorescence   The property of absorbing light and then reemitting light of a wavelength unique to the fluorescing compound.

Food Vacuole   An intracellular membrane-bound compartment containing an engulfed particle of food.

Free Energy   The total chemical and physical energy of a substance available to do work.

Frond   The leaf of a fern; any large divided leaf.

Fruit   A mature, ripened ovary or group of ovaries.

Fungi   A eukaryotic, heterotrophic, haploid organism with chitinous or cellulose cell walls; usually saprophytic or parasitic.

G

Gamete   Haploid reproductive cell whose nucleus fuses with that of another gamete of the opposite sex to form a zygote.

Gametogenesis   The formation and differentiation of gametes.

Gametophyte   The haploid gamete-producing phase of plants characterized by alternation of generations.

Gap$_1$ Phase   The stage of interphase of the cell cycle occurring after mitosis and including a variety of metabolic processes.

Gap$_2$ Phase   The stage of interphase of the cell cycle occurring just before mitosis and including the synthesis of compounds necessary for mitosis.

Gene   A basic unit of heredity; a sequence of DNA nucleotides on a chromosome.

Genotype   The total set of genes present in the cells of an organism.

Gibberellic Acid   A plant hormone that elongates plant stems among other effects.

Glycolysis   The anaerobic breakdown of glucose; the enzyme-catalyzed breakdown of glucose to two molecules of pyruvate with the net liberation of two molecules of ATP.

Gravitropism   Growth response to gravity in plants; formerly called geotropism.

Growth   Increase in size of an organism caused by increase in the amount of protoplasm.

H

Haploid   Having a single set of chromosomes.

Hardy-Weinberg Law   A mathematical description of the relative frequencies of alleles in a population subject to Mendelian segregation.

Haustorium   A protuberance of a fungal hypha that functions as a penetrating and absorbing structure.

Heterocyst   A large, thick-walled cell that forms under appropriate conditions in the filaments of certain cyanobacteria; nitrogen fixation occurs in heterocysts.

Heterosporous   Having two kinds of spores.

Heterotrophic   Deriving energy from organic molecules made by other organisms.

Heterozygote   A diploid individual that carries two different alleles on homologous chromosomes at one or more genetic loci.

Homologous   Having structure or function indicative of a common evolutionary origin.

Homologous Chromosomes   In diploid or polyploid organisms, the chromosomes that have the same gene loci.

Homosporous   Having one kind of spore.

Homozygote   A diploid individual that carries identical alleles at one or more genetic loci.

Host Organism   An organism harboring a parasite.

Hydrolysis   The breaking of a chemical bond by the addition of a water molecule.

Hydrophilic   Condition of being attracted by water, "water-loving"; describes most polar molecules.

Hydrophobic   Condition of being repulsed by water, "water-fearing"; describes most nonpolar molecules.

Hypertonic   Having a higher concentration of solutes than the solution being compared.

Hypha   A filament of cells in a fungus.

Hypotonic   Having a lower concentration of solutes than the solution being compared.

I

Indicator   A chemical whose color change indicates a change in conditions.

Interphase   The metabolically active stage of the cell cycle when the events of mitosis are not occurring.

Isogamy   Reproduction involving gametes that are similar in size and appearance.

Isotonic   Having the same concentration of solutes as the solution being compared.

L

Lenticels   Spongy areas in the cork surfaces of stems, roots, and other plant parts that allow interchange of gases between internal tissues and the atmosphere through the periderm.

Lichen   An organism composed of fungi and algae (or blue-green bacteria) living symbiotically.

Lysis   Rupture of a cell by internal hydrostatic pressure.

M

Macronucleus   A large membrane-bound nucleus common in ciliate protozoans.

Macronutrients   Inorganic chemical elements required in large amounts for plant growth, such as nitrogen, potassium, calcium, phosphorus, magnesium, and sulfur.

Megaspore   In heterosporous plants, a haploid (1N) spore that develops into a female gametophyte; megaspores are usually larger than microspores.

Meiosis   In the sexual reproduction of eukaryotes, the nuclear division in which a diploid cell gives rise to four haploid nuclei; the haploid nuclei often are gametes.

Meristem   In plants, a region of actively reproducing cells forming new tissues.

Metaphase   The second phase of mitosis characterized by alignment of chromosomes on the spindle apparatus.

Micronucleus   A small membrane-bound nucleus common in ciliate protozoans.

Micronutrient   A mineral required in only minute amounts for plant growth, such as iron, chlorine, copper, manganese, zinc, molybdenum, and boron.

Microspore   In plants, a spore that develops into a male gametophyte; in seed plants, it develops into a pollen grain.

Mitosis   Replication of a eukaryotic nucleus and its genetic material.

Monocotyledon   A class of flowering plants generally characterized as having one cotyledon.

Morphogenesis   The development of form.

Mutation   A permanent change in the genetic code of a cell's DNA.

Mycelium   The mass of hyphae that composes a fungal organism.

N

Natural Selection   The differential reproduction of genotypes caused by factors in the environment.

Nitrogen Fixation   The incorporation of atmospheric nitrogen into organic nitrogen compounds, a process that can be carried out only by certain microorganisms.

Nodule   Swelling of the roots of legumes and other plants inhabited by nitrogen-fixing bacteria.

Nonpolar Molecules   Molecules with no areas of localized positive or negative charge.

O

Oogamy   Sexual reproduction in which one gamete (sperm) is small and motile, and the other gamete (egg) is larger and nonmotile.

Osmosis   Movement of water through semipermeable biological membranes.

Ovary   In flowering plants, the enlarged basal portion of the carpel, which contains the ovule(s); the ovary matures to become the fruit.

Ovule   Structure containing female gametophyte (with egg) surrounded by the nucellus and integuments; develops into a seed.

P

Parasite   An organism that lives and feeds on another living organism.

Petiole   The stalk of a leaf.

pH   A measure of the relative concentration of hydrogen ions in a solution; equal to the negative logarithm of the hydrogen ion concentration. Values range from 0 to 14. The lower values indicate greater acidity.

Phenotype   The realized expression of a genotype.

Phloem   In vascular plants, a food-conducting tissue.

Phospholipid Bilayer   A unit of two parallel layers of phospholipid molecules that constitutes cellular membranes.

Photosynthesis   The utilization of light energy to create chemical bonds; the synthesis of organic components from carbon dioxide and water, using chemical energy (ATP) and reducing power (NADPH).

Phototropism   In plants, a growth response to a light stimulus.

Phycobilins   A group of water-soluble accessory pigments, including phycocyanins and phycoerythrins, which occur in red algae and cyanobacteria.

Phytochrome   A plant pigment associated with the absorption of light; photoreceptor for red to far-red light; involved in a number of timing processes, such as flowering, dormancy, leaf formation, and seed germination.

Pigment   A light-absorbing substance.

Plant   A eukaryotic, multicellular, autotrophic organism; usually with cell walls of cellulose.

Plasmodium   A vegetative stage in the life cycle of some slime molds in which individual cells are not distinguishable.

Plastid   Organelle in plants and algae that is the site of photosynthesis.

Polar Molecules   Molecules with a spatial separation of electrical charge.

Pollination   Transfer of pollen to receptive surface.

Primary Growth   Growth or new tissue in plants laid down by the apical meristem.

Product   In an enzymatic reaction, the molecule resulting from the action of an enzyme on its substrate.

Prokaryotic   Term applied to cells lacking nuclei.

Prophase   The first stage of mitosis characterized by condensed chromosomes.

Protist   A eukaryotic, microscopic, unicellular organism belonging to the Kingdom Protista.

Protoplasm   All living material composing cells and living structures.

R

Resolution   The ability to visually distinguish two points as separate.

Respiration   The cellular breakdown of sugar and other foods to release energy held in the carbon bonds and to capture it in the usable form of ATP.

Rhizoid   Slender rootlike anchoring structure in some fungi, algae, and plant gametophytes.

Rhizome   In vascular plants, a usually more or less horizontal underground stem; may function in vegetative reproduction.

Root Hair   In vascular plants, a tubular outgrowth of the epidermal cells of the root just behind its apex; most water enters through the root hairs.

S

Saprophyte   An organism that feeds on dead organic matter.

Secondary Growth   Growth or new tissue in plants laid down by non-apical meristems.

Seed   Mature ovule consisting of a developing embryo, a food supply, and a protective seed coat.

Septa   Cross walls or partitions that distinguish cells or compartmentalize larger structures.

Sieve Tube   A column of cells called elements arranged end to end that conduct food from element to element throughout the plant.

Spectrophotometry   A technique for measuring the absorbance and transmittance of electromagnetic radiation by dissolved substances.

Spermatogenesis   The formation and differentiation of sperm cells.

Sperm Cell   A male gamete.

Spindle Apparatus   The assembly that carries out the separation of chromosomes during cell division; composed of microtubules and assembled during prophase at the equator of the dividing cell.

Spindle Fibers   A group of microtubules that together make up the spindle apparatus.

Sporangium   A structure bearing spores.

Spore   Reproductive cell capable of developing into an adult without fusing with another cell.

Sporophyte   The spore-producing diploid (2N) phase in the life cycle of a plant having alternation of generations.

Standard Curve   A graph of a chemical's absorbance of a particular wavelength of light versus the chemical's concentration; constructed using chemical samples of known (standard) concentration.

Starch   A polymer of glucose used as an energy storage molecule by plants.

Stolon   A stem or fungal hypha that grows horizontally and gives rise to upright leaves or sporangiophores.

Stomata   In plants, a minute opening bordered by guard cells in the epidermis of leaves and stems; water passes out of a plant mainly through the stomata, and carbon dioxide passes in chiefly by the same pathway.

Substrate   In an enzymatic reaction, the molecules reacting with the enzyme.

Symbiosis   Two or more dissimilar organisms living together in close association.

Synapsis   The point-by-point alignment (pairing) of homologous chromosomes that occurs before the first meiotic division; crossing-over occurs during synapsis.

Syngamy   The process by which two haploid cells fuse to form a diploid zygote; fertilization.

Synthesis Phase   The stage of interphase of the cell cycle in which chromosomal DNA is replicated.

T

Telophase   The fourth and final stage of mitosis characterized by the formation of two new, membrane-bound nuclei having diffuse chromosomes.

Thallus   Body of a plant lacking true roots, stems, and leaves.

Tonoplast   Membrane surrounding a vacuole.

Tracheid   In vascular plants, an elongated, thick-walled conducting and supporting

2-271

cell of xylem. Tracheids, which are dead when functional, have tapered ends and pitted walls without perforations.

Transformation The transfer of free DNA from one organism to another.

Transpiration The loss of water vapor by plant parts.

Tropism A response to an external stimulus.

Vector In genetics, any virus or plasmid DNA into which a gene is integrated and subsequently transferred into a cell; in

disease, any agent that carries or transmits a pathogenic microorganism from one host to another host.

Wet Mount A preparation of fresh material for examination with a microscope.

X

Xylem In vascular plants, water-conducting tissue.

Z

Zygospore A thick-walled resistant spore that develops from a zygote, resulting from the fusion of isogametes.

Zygote The diploid (2N) cell resulting from fusion of the male and female gametes.

credits

PHOTOGRAPHS

EXERCISE 1

1.1: Courtesy of Carl Zeis, Inc., Thornwood, NY; **1.2:** Darrell Vodopich; **1.6:** Courtesy of Leica, Inc.

EXERCISE 3

3.2A, 3.5, 3.6, 3.7, 3.9, 3.10: Courtesy of Edvotek Inc.; **3.11E:** Courtesy of George Kantor

EXERCISE 4

4.1A: © David M. Phillips/Visuals Unlimited; **4.2:** Courtesy of E.H. Newcomb & T.D. Pugh, University of Wisconsin; **4.3A:** Courtesy of Kenneth Miller, Brown University; **4.4B:** © Don W. Fawcett/Visuals Unlimited; **4.8:** © John D. Cunningham/Visuals Unlimited; **4.9:** © Dr. Jeremy Burgess/SPL/Photo Researchers, Inc.; **4.10:** Courtesy of Barbara Hull

EXERCISE 5

5.3: Photo by J. David Robertson from Charles Flickinger, *Medical Cellular Biology*, W.B. Saunders, 1979.

EXERCISE 11

11.2: © Biophoto Associates/Photo Researchers, Inc.; **11.4A:** © B.A. Palevitz & E.H. Newcomb/BPS/ Tom Stack & Associates; **11.5A:** © Ed Reschke; **11.5B:** © Biodisc, Inc.; **11.5 C-E:** © Ed Reschke

EXERCISE 13

13.1: Courtesy of Ulrich Laemmli; **13.3:** Courtesy of Stanley N. Cohen, Stanford University

EXERCISE 16

16.2, 16.3: Randy Moore; **16.4D:** Courtesy of Wilfred Cote, SUNY, College of Environmental Forestry

EXERCISE 17

17.1, 17.3: Randy Moore; **17.4:** © BioPhot; **17.5:** © Ed Reschke; **17.6, 17.7, 17.8:** Randy Moore

EXERCISE 19

19.2: © Carolina Biological/Phototake, NYC

EXERCISE 20

20.3: © Sylvan Wittwer/Visuals Unlimited

EXERCISE 21

21.2 A-B: © BioPhot; **21.3 A-B:** © Professor Malcolm B. Wilkins, Botany Department, Glasgow University

EXERCISE 22

22.1: Randy Moore

EXERCISE 23

23.2, 23.3, 23.4, 23.5, 23.6: © Carolina Biological/Phototake, NYC

EXERCISE 24

24.2: © E.C.S. Chan/Visuals Unlimited

EXERCISE 25

25.5: Darrell Vodopich; **25.6A:** © David Phillips/Visuals Unlimited; **25.6B:** © Biophoto Associates/Photo Researchers, Inc.; **25.7:** © David M. Phillips/Visuals Unlimited; **25.9, 25.11, 25.12, 25.13, 25.14 A-C:** Darrell Vodopich

EXERCISE 26

26.6: Courtesy of J.D. Pickett-Heaps

EXERCISE 27

27.3, 27.5, 27.6, 27.8, 27.9, 27.10: Randy Moore

EXERCISE 28

28.2: Randy Moore; **28.3:** Darrell Vodopich; **28.4:** Randy Moore; **28.5, 28.6:** Darrell Vodopich; **28.7, 28.8:** Randy Moore

EXERCISE 29

29.2: © Ed Reschke; **29.3:** Darrell Vodopich; **29.5:** Randy Moore; **29.7:** © George J. Wilder/Visuals Unlimited; **29.8 A-B:** © Robert & Linda Mitchell; **29.9:** Darrell Vodopich

EXERCISE 18

18.3: From Peter H. Raven and George B. Johnson, *Biology*, 3d ed. Copyright © 1995 The McGraw-Hill Companies, Inc. All Rights Reserved. Reprinted by permission. **18.4:** From Kingsley R. Stern, *Introductory Plant Biology*, 6th ed. Copyright © 1994 The McGraw-Hill Companies, Inc. All Rights Reserved. Reprinted by permission.

EXERCISE 19

19.1: From Peter H. Raven and George B. Johnson, *Biology*, 3d ed. Copyright © 1995 The McGraw-Hill Companies, Inc. All Rights Reserved. Reprinted by permission.

EXERCISE 20

20.2: From Peter H. Raven and George B. Johnson, *Understanding Biology,* 3d ed. Copyright © 1995 The McGraw-Hill Companies, Inc. All Rights Reserved. Reprinted by permission.

EXERCISE 24

24.1: From Randy Moore and Darrell Vodopich, *General Biology Laboratory Manual.* Copyright © 1991 The McGraw-Hill Companies, Inc. All Rights Reserved. Reprinted by permission. **24.3:** From Randy Moore and Darrell Vodopich, *General Biology Laboratory Manual.* Copyright © 1991 The McGraw-Hill Companies, Inc. All Rights Reserved. Reprinted by permission. **24.4:** From Randy Moore and Darrell Vodopich, *General Biology Laboratory Manual.* Copyright © 1991 The McGraw-Hill Companies, Inc. All Rights Reserved. Reprinted by permission. **24.5:** From Randy Moore and Darrell Vodopich, *General Biology Laboratory Manual.* Copyright © 1991 The McGraw-Hill Companies, Inc. All Rights Reserved. Reprinted by permission.

EXERCISE 25

25.4: From Peter H. Raven and George B. Johnson, *Biology*, 2d ed. Copyright © 1989 The McGraw-Hill Companies, Inc. All Rights Reserved. Reprinted by permission. **25.8:** From Peter H. Raven and George B. Johnson, *Biology,* 3d ed. Copyright © 1995 The McGraw-Hill Companies, Inc. All Rights Reserved. Reprinted by permission. **25.10:** From Peter H. Raven and George B. Johnson, *Biology,* 2d ed. Copyright © 1989 The McGraw-Hill Companies, Inc. All Rights Reserved. Reprinted by permission.

EXERCISE 26

26.1: From Peter H. Raven and George B. Johnson, *Biology*, 2d ed. Copyright © 1989 The McGraw-Hill Companies, Inc. All Rights Reserved. Reprinted by permission. **26.4:** From Randy Moore and Darrell Vodopich, *General Biology Laboratory Manual.* Copyright © 1991 The McGraw-Hill Companies, Inc. All Rights Reserved. Reprinted by permission. **26.8:** From Peter H. Raven and George B. Johnson, *Biology*, 4th ed. Copyright © 1996 The McGraw-Hill Companies, Inc. All Rights Reserved. Reprinted by permission. **26.7:** From Peter H. Raven and George B. Johnson, *Biology*, 4th ed. Copyright © 1996 The McGraw-Hill Companies, Inc. All Rights Reserved. Reprinted by permission.

EXERCISE 27

27.1: From Peter H. Raven and George B. Johnson, *Understanding Biology*, 3d ed. Copyright © 1995 The McGraw-Hill Companies, Inc. All Rights Reserved. Reprinted by permission. **27.2:** From Randy Moore and Darrell Vodopich, *General Biology Laboratory Manual.* Copyright © 1991 The McGraw-Hill Companies, Inc. All Rights Reserved. Reprinted by permission. **27.4:** From Randy Moore and Darrell Vodopich, *General Biology Laboratory Manual.* Copyright © 1991 The McGraw-Hill Companies, Inc. All Rights Reserved. Reprinted by permission. **27.7:** From Peter H. Raven and George B. Johnson, *Biology*, 3d ed. Copyright © 1995 The McGraw-Hill Companies, Inc. All Rights Reserved. Reprinted by permission.

EXERCISE 28

28.1: From Peter H. Raven and George B. Johnson, *Biology*, 3d ed. Copyright © 1995 The McGraw-Hill Companies, Inc. All Rights Reserved. Reprinted by permission.

EXERCISE 29

29.1: From Peter H. Raven and George B. Johnson, *Understanding Biology*, 3d ed. Copyright © 1995 The McGraw-Hill Companies, Inc. All Rights Reserved. Reprinted by permission. **29.4:** From Peter H. Raven and George B. Johnson, *Biology*, 2d ed. Copyright © 1989 The McGraw-Hill Companies, Inc. All Rights Reserved. Reprinted by permission.

EXERCISE 30

30.1: From Peter H. Raven and George B. Johnson, *Biology*, 3d ed. Copyright © 1995 The McGraw-Hill Companies, Inc. All Rights Reserved. Reprinted by permission. **30.2:** From Peter H. Raven and George B. Johnson, *Understanding Biology*, 3d ed. Copyright © 1995 The McGraw-Hill Companies, Inc. All Rights Reserved. Reprinted by permission. **30.9:** From Peter H. Raven and George B. Johnson, *Biology*, 3d ed. Copyright © 1995 The McGraw-Hill Companies, Inc. All Rights Reserved. Reprinted by permission.

PLATE FIGURES

1, 2, 4, 6, 8, 9, 10, 11, 14: From Peter H. Raven and George B. Johnson, *Biology*, 4th ed. Copyright © 1996 The McGraw-Hill Companies, Inc. All Rights Reserved. Reprinted by permission.

3, 5, 7, 9, 12, 13: From Peter H. Raven and George B. Johnson, *Understanding Biology*, 3rd ed. Copyright © 1995 The McGraw-Hill Companies, Inc. All Rights Reserved. Reprinted by permission.

15: From Peter H. Raven and George B. Johnson, *Biology*, 3rd. ed. Copyright © 1995 The McGraw-Hill Companies, Inc. All Rights Reserved. Reprinted by permission.